Principles and Practices of Electrical Safety Engineering

Ensuring Protection in Electrical Systems

Published 2025 by River Publishers
River Publishers
Alsbjergvej 10, 9260 Gistrup, Denmark
www.riverpublishers.com

Distributed exclusively by Routledge
605 Third Avenue, New York, NY 10017, USA
4 Park Square, Milton Park, Abingdon, Oxon OX14 4RN

Principles and Practices of Electrical Safety Engineering / by Massimo Mitolo.

Routledge is an imprint of the Taylor & Francis Group, an informa business

ISBN 978-87-7004-756-2 (paperback)

ISBN 978-87-7004-758-6 (online)

ISBN 978-87-7004-757-9 (ebook master)

A Publication in the River Publishers Series in Rapids

Principles and Practices of Electrical Safety Engineering
Ensuring Protection in Electrical Systems

Massimo Mitolo

Irvine Valley College, California, USA

River Publishers

Routledge
Taylor & Francis Group

NEW YORK AND LONDON

Dedicated to my Father, Dean and Professor Domenico Mitolo.
First among the men.
Brilliant physicist; brilliant mathematician.
He lived heroically; and he still does.
In my heart.

Contents

Contents

Preface

The field of electrical safety engineering is fundamental to the design, operation, and maintenance of electrical systems, ensuring the protection of individuals from electrical hazards. This book aims to provide a comprehensive understanding of the principles and practices essential to achieving this goal.

Throughout the chapters, readers will find numerous explanatory figures that illustrate key concepts and processes. These figures serve as a valuable visual aid, enhancing comprehension and making complex ideas more accessible. Whether depicting the intricate workings of safety devices or the layout of protective measures, these illustrations are designed to support and clarify the text.

Electrical safety is not merely a set of guidelines to follow; it is an integral aspect of engineering design. The concept of "safety-by-design" is emphasized throughout this book, highlighting the importance of integrating safety measures from the earliest stages of development. By proactively identifying and mitigating potential hazards, engineers can create safer environments and reduce the risk of incidents.

This book is dedicated to the importance of electrical safety engineering, detailing various protective measures and their implementation. Topics such as risk assessment, fault protection, and the reliability of safety systems are thoroughly examined. These discussions underscore the critical role that meticulous planning and robust design play in preventing electrical incidents.

The insights and knowledge presented in this book are intended to equip engineers, technicians, and students with the tools necessary to ensure electrical safety. By fostering a deep understanding of both the theoretical and practical aspects of electrical safety engineering, this book aims to contribute to the advancement of safer electrical practices worldwide.

About the Author

Dr. Massimo Mitolo, a distinguished scholar and scientist, has been bestowed a Knighthood in the *Order of Merit of the Italian Republic* in acknowledgment of his exceptional contributions to scientific endeavors that have brought great honor to the nation. He is renowned for his remarkable achievements in the field of electrical engineering.

Sir Massimo earned his Ph.D. in Electrical Engineering from the University of Napoli "Federico II" in Italy. His unwavering dedication and significant impact on the field have led to his recognition as a Fellow of IEEE "for contributions to the electrical safety of low-voltage systems". Furthermore, he holds the distinguished title of Fellow from the Institution of Engineering and Technology (IET) in London, United Kingdom, and is a member of the IEEE-HKN Honor Society. Additionally, he is a registered Professional Engineer in both the state of California and Italy.

Presently, Dr. Mitolo serves as a Full Professor of Electrical Engineering at Irvine Valley College in California. In addition to his academic responsibilities, he is a senior consultant specializing in the domains of failure analysis and electrical safety. His extensive research and industrial experience revolve around the comprehensive analysis and grounding of power systems, as well as electrical safety engineering.

Dr. Mitolo's expertise is reflected in his publication record, encompassing more than 190 journal papers, as well as the authorship of several influential books. Noteworthy titles authored by him include "Electrical Safety of Low-Voltage Systems" (McGraw-Hill, 2009), "Laboratory Manual for Introduction to Electronics: A Basic Approach" (Pearson, 2013), "Analysis of Grounding and Bonding Systems" (CRC Press, 2020), "Electrical Safety Engineering of Renewable Energy Systems" (IEEE Wiley, 2021), "Smart and Power Grid Systems: Design Challenges and Paradigms" (River Publishers 2022), and "Simulation-based Labs for Circuit analysis." (River Publishers, 2024).

His scholarly endeavors have garnered significant recognition, culminating in his inclusion in the 2020, 2021, 2022 and 2023 World's Top 2% Most-cited Scientists List, as compiled by Stanford University.

Within the Industrial and Commercial Power Systems Department of the IEEE Industry Applications Society (IAS), Dr. Mitolo actively engages in various committees and working groups, demonstrating his commitment to advancing the field and fostering collaborative efforts.

Acknowledging his achievements, Dr. Mitolo has been the recipient of numerous prestigious accolades throughout his career. Notably, he has been honored with the IEEE Region 6 Outstanding Engineer Award and has garnered nine Best Paper Awards for his exceptional scholarly contributions. Furthermore, he has received recognitions such as the IEEE Ralph H. Lee I&CPS Department Prize Award, the IEEE I&CPS Department Achievement Award, and the James E. Ballinger Engineer of the Year Award from the Orange County Engineering Council in California.

Principles of Electrical Safety Engineering

1.1 Introduction

The primary objective of electrical safety engineering is to prevent incidents that can have severe consequences for individuals. These incidents often occur when an unsafe act intersects with an unsafe situation, stemming from factors such as environmental conditions, human error, and equipment faults. Ensuring the safe operation of electrical systems involves designing, implementing, and maintaining safety measures to protect individuals and property from electrical risks, including shocks, electrocution, fires, and explosions.

At the heart of preventing electrical incidents is the process of risk assessment. During the design phase, engineers identify potential hazards and evaluate the risks associated with electrical systems. This process underscores that electrical safety cannot depend solely on a set of prudent actions around energized parts; it must be ingrained in the design phase. The concept of safety-by-design involves anticipating and eliminating hazards before they can cause harm. Through this proactive strategy, protective measures are integrated into electrical systems and equipment from the outset. By embedding safety features into the design, hazards are minimized or eliminated, creating safer environments and reducing the likelihood of injuries.

Proper risk assessment and thoughtful electrical design work hand in hand to establish a robust framework for electrical safety. This integrated approach enables engineers to foresee potential hazards and address them effectively. It highlights the importance of comprehensive planning in electrical safety, ultimately ensuring the protection and well-being of individuals.

1.1.1 Factors contributing to electrical incidents

Several factors contribute to electrical incidents, each impacting safety in distinct ways. Understanding and managing these factors is crucial for maintaining safe electrical systems.

The physical and environmental conditions surrounding electrical equipment significantly impact safety. Factors such as temperature, humidity, and the presence of corrosive substances can affect the integrity of electrical installations. These external conditions must be carefully managed to ensure the longevity and reliability of electrical systems. For example, high humidity can lead to moisture accumulation, increasing the risk of short circuits, while corrosive environments can degrade insulation and other protective measures.

Human error is a common cause of electrical accidents. This can stem from inadequate training, improper maintenance, or incorrect operation of electrical equipment. Ensuring that personnel are well-trained and follow proper procedures is crucial for minimizing these risks. Regular training programs and strict adherence to safety protocols help reduce the likelihood of mistakes that could lead to dangerous situations.

The design, quality, and condition of electrical equipment are critical to safety. Failures such as insulation breakdowns or mechanical faults can expose individuals to electric shock. Using high-quality materials and conducting rigorous testing are essential to prevent such failures. Routine inspections and maintenance are also necessary to ensure that equipment remains in good working condition.

Certain incidents or conditions, such as power surges, short circuits, or natural disasters, can trigger electrical incidents regardless of human behavior. These events can cause significant damage and pose serious safety risks. For instance, power surges can damage electronic components, while natural disasters like floods or earthquakes can disrupt electrical installations and create hazardous situations.

The primary concern in electrical safety engineering is the prevention of injuries, which can range from minor shocks to severe burns or fatalities. A comprehensive approach to safety is essential, encompassing environmental management, human factors, equipment design, and incident response. By addressing each of these areas, it is possible to protect individuals and maintain safe electrical systems.

The interplay between environmental conditions, human factors, equipment design, and specific incidents necessitates a holistic approach to electrical safety. This comprehensive strategy is vital to preventing injuries and ensuring the safe operation of electrical systems.

1.1.2 Consequences of electrical failures

When electrical equipment fails, the consequences can be severe and multi-faceted. The potential loss of human life underscores the necessity of rigorous safety measures. Economic loss is a significant concern, as damage to structures and contents can lead to substantial financial setbacks. Service disruption is another critical issue, as interruptions to public services can impact communities and businesses, causing widespread inconvenience. Additionally, electrical fires in museums or historical sites can result in the irretrievable loss of cultural assets, representing a profound cultural heritage loss. These consequences highlight the need for robust electrical safety protocols and preventive measures to mitigate risks and protect both people and property.

1.2 Protective Measures and Electrical Safety

Electrical safety is achieved by integrating standard protective measures (e.g., basic insulation of live parts) into equipment, aimed at reducing the risk of electric shock to acceptable levels. Protective measures turn the electrical equipment into a system whose safety depends on the reliability of each measure. Since protective measures have a non-zero failure rate, their occasional failure is inevitable[1]. Electrical safety, up to a specified time t, can be defined as the probability that the equipment will continue to function properly without causing dangerous voltage exposure despite the occurrence of random faults.

---

[1]M. Mitolo, "Is it Possible to Calculate Safety: Safety and Risk Analysis of Standard Protective Measures Against Electric Shock," in IEEE Industry Applications Magazine, vol. 15, no. 3, pp. 31-35, May-June 2009.

To clarify the concept, let us consider a population of N nominally identical electrical pieces of equipment that function under similar operating conditions for the time t. These objects may be exposed to a specific fault (e.g., failure of basic insulation) that can create an unsafe condition for persons (e.g., electric shock). We mathematically define the safety, $S(t)$, of any of these N objects against the unsafe condition with Equation (1.1).

$$S(t) = \frac{R(t)}{N} = e^{-\lambda t}, \tag{1.1}$$

where $R(t)$ is the *reliability* function, representing the number of identical items that survived the specific fault by the time t, and λ is their *failure rate*, assumed constant. $S(t)$ is the probability that within a specified time an adverse event will not occur and uses the negative exponential distribution to represent its time evolution.

In discrete terms, the failure rate λ is defined as the ratio of the number of objects that fail within a population to the total time of operation expended by all the objects during a particular measurement interval under stated conditions. For example, consider ten identical transformers whose basic insulation is monitored for 1000 hours under test conditions. At the end of the test, the basic insulation of three transformers failed after 750, 900, and 950 hours, while seven transformers survived with the basic insulation intact. The failure rate of the basic insulation of those transformers can be calculated as follows:

$$\lambda = \frac{\text{Number of failures}}{\text{total time}} = \frac{3}{750 + 900 + 950 + 7 \cdot 1000}$$
$$= 0.3125 \cdot 10^{-3} \text{ failures/hours} \tag{1.2}$$

From Equation (1.1), it is evident that, as time progresses, $S(t)$ exponentially decays to zero, indicating that incidents are probable over a very long period. Safety is at its maximum (i.e., $S(t) = 1$) when the equipment is not energized ($t = 0$), operates at non-dangerous voltages, or if $\lambda = 0$, which is impossible in practice.

Safety is also zero in situations when a fault is not necessary to create a specific hazardous condition for individuals. This scenario is inherently dangerous regardless of whether a failure occurs and is often associated with an infinite failure rate because the risk is constantly present. Consider, for instance, a bare high-voltage electrical conductor within easy reach. This represents a zero-safety scenario because the risk of severe injury or fatality is present at all times and is not contingent on any system failure.

This approach allows for a quantitative assessment of safety, providing a metric to evaluate the likelihood of adverse events and the overall performance of the equipment. By understanding and monitoring these metrics, engineers can implement more effective maintenance strategies, design improvements, and safety protocols to enhance the reliability and safety of electrical systems.

1.3 Safety of Electrical Systems

When multiple protective measures against electric shock are deployed simultaneously, their combined effectiveness must be considered to calculate the overall safety of the system. The configuration of these measures determines the system's reliability and safety characteristics.

In a *series* protection system, all *m* protective measures must function correctly at the same time to ensure safety. This type of system is highly dependent on the reliability of each individual measure, as the failure of a single component can lead to a failure of the entire safety.

Example of series protection against electric shock is the protective measure through insulation that depends on both *creepage* distance (i.e., surface distance) and *clearance* distance (i.e., air distance). Here, the insulation's effectiveness relies on maintaining specific creepage and clearance distances between live parts and if either distance is compromised, the insulation may fail, leading to a potential safety hazard. It is clear that the overall safety of a series system against a risk decreases as the number of protective measures increases, and its value is less than the safety of the least reliable component. To improve the safety of a serial system, it is therefore advisable to enhance the safety of the least reliable component.

In a *parallel* (or *redundant*) arrangement of protective measures, the system remains safe as long as at least one of the *m* protective measures is functioning correctly. For the system's safety to be compromised, all protective measures must fail. This configuration enhances the overall reliability and safety, as it provides multiple layers of protection. Redundancy ensures that even if one or more measures fail, others can still maintain the system's safety. An example of a parallel protection system is double insulated equipment (also referred to as Class II equipment), which incorporates a supplementary layer of insulation in addition to the basic insulation. The basic insulation provides the initial barrier against electrical hazards and the supplementary insulation is an additional layer that acts as a backup in case the basic insulation fails. The two insulations are physically arranged so that they cannot be subject to the same deteriorating factors to the same degree.

Safety for serial and parallel protective measures can be respectively evaluated through Equations (1.3) and (1.4).

$$S_S(t) = S_1(t) \cdot ... \cdot S_{n-1}(t) \cdot S_n(t) \tag{1.3}$$

$$S_P(t) = 1 - [1 - S_1(t)] \cdot ... \cdot [1 - S_n(t)]. \tag{1.4}$$

The term *danger* refers to the probability that an adverse event, such as hazardous potential on the equipment enclosure, will occur within a specified time frame. This probability is represented by the quantity $(1 - S(t))$.

1.4 The Electrical Risk

Electric contact can occur in two main forms: *direct* contact and *indirect* contact. Direct contact refers to contact with a part that is normally live, such as an exposed conductor or a terminal. Indirect contact, on the other hand, involves touching exposed-conductive parts of equipment that have become live due to a failure in the basic insulation. Both types of contact pose significant risks to individuals, potentially leading to physical injury or death. Selecting appropriate protective measures against electric shock requires a thorough evaluation of the actual electrical risk *r(t)* involved.

The risk at the time t can be estimated by Equation (1.5).

$$r(t) = [1 - S(t)] \cdot k(t) \cdot u(t). \tag{1.5}$$

The quantity $1 - S(t)$ represents the probability that dangerous potentials are present on the equipment enclosure or component. The term $k(t)$ denotes the probability that persons will touch the energized part. For portable electric equipment, intended to be held in the hand during normal use, $k(t) = 1$; in this case, the presence of voltages under fault conditions on the equipment can be properly prevented by double insulation. For fixed electric equipment, which is fastened to a support or otherwise secured in a specific location, $k(t) < 1$, and the protective measure of automatic disconnection of the supply in conjunction with protective bonding is generally adopted. The factor $u(t)$ indicates the probability of a person being exposed to touch voltages exceeding conventional safe limits or for unsafe durations. Consequently, $u(t)$ represents the probability of harm to persons.

The conventional safe voltage limits differ based on the type of contact. For indirect contact, in normal dry conditions such limits are 50 V a.c. or 120 V d.c.

For direct contact with live parts, the voltage limits are 25 V a.c. or 60 V d.c.[2] The reason behind the two different sets of voltage limits is explained in reference to Figure 1.1.

Figure 1.1: (a) Direct contact; (b) indirect contact.

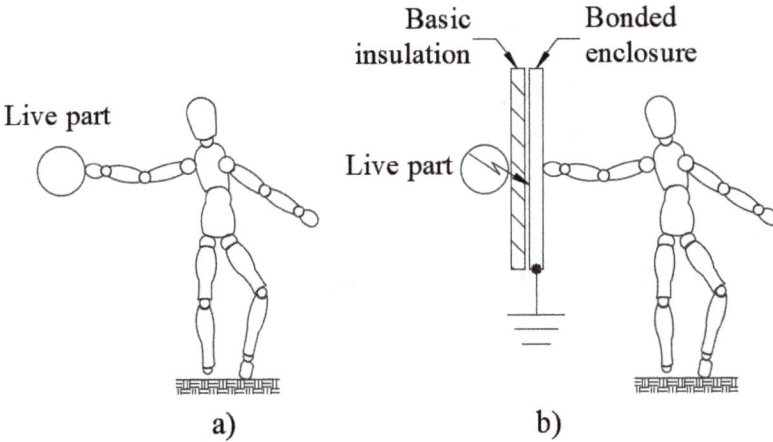

When an individual comes into contact with a live part, a direct contact occurs (Figure 1.1 a). Conversely, if they touch an exposed-conductive part of equipment that has become live due to a fault, an indirect contact takes place (Figure 1.1 b). The harm $u(t)$ resulting from interaction with an electrical component or equipment is determined by the magnitude of the touch voltage and is consistent for both contact types. Similarly, the probability $k(t)$ that a person will touch the electrical component or equipment is the same for both direct and indirect contact. However, the probability $1 - S(t)$ that dangerous potentials are present varies significantly between the two contact types.

In the case of direct contact, this probability is one, as by definition the part is live and entirely exposed to touch. Conversely, for indirect contact, the probability of equipment to remain energized due to basic insulation failure is low. The safety measure provided by the equipotential bonding and the protective device has the goal of promptly disconnecting the supply even before an individual can touch the equipment. Consequently, the risk $r(t)$ associated with direct contact is greater than that associated with indirect contact, given the same touch voltage magnitude. To balance the risks, the conventional safe voltage limits for direct contact are, therefore, lower.

---

[2]IEC 60364-4-41:2005, "Low-voltage electrical installations – Part 4-41: Protection for safety – Protection against electric shock."

1.5 Class I Equipment

Consider a piece of equipment with basic insulation only, which fails (e.g., a damaged wire) (Figure 1.2).

Figure 1.2: Faulty equipment with basic insulation only.

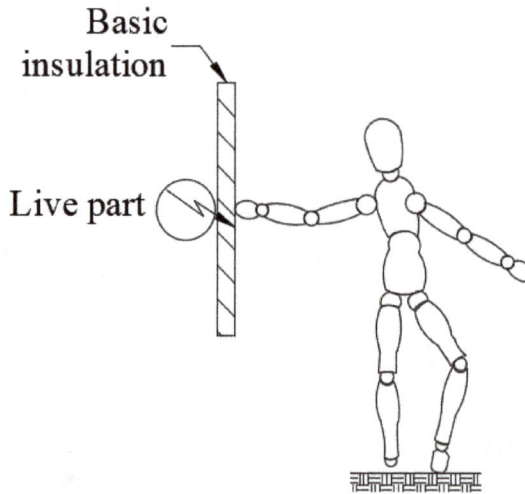

For this adverse event to result in harm, an individual must make contact precisely at the fault point, a very low-probability event. This scenario changes if the insulated live part is enclosed within a conductive frame. In this case, upon failure of the basic insulation, any point of contact with the enclosure could result in harm.

Enclosing an insulated live part within a conductive frame does not alter the failure rate λ_B of the basic insulation, or the probability of harm $u(t)$, but increases the probability $k(t)$ that individuals may be exposed to a touch potential. The electrical risk involved is therefore increased and must be mitigated. This is generally (but not necessarily, as discussed in the next chapters) obtained by implementing the protective measure of automatic disconnection of the supply, which consists of protective device and equipotential bonding. Equipment that provides basic insulation and includes a terminal for protective equipotential bonding for fault protection is classified as Class I (Figure 1.1 b).

A parallel protective measure protects the Class I equipment against indirect contact. Even if the basic insulation fails, the protective device can clear the fault thanks to the bonding connection. Conversely, the failure of the bonding connection and/or the protective device does not expose individuals to hazardous potential because of the basic insulation. However, the equipotential bonding connection and the protective device is a series measure, whose expression for the safety $S_1(t)$ is given by Equation (1.6):

$$S_1(t) = S_{EQB}(t) \cdot S_{PD}(t) = e^{-(\lambda_{EQB} + \lambda_{PD})t} = e^{-\lambda_1 t} \tag{1.6}$$

where λ_{EQB} and λ_{PD} are the failure rates of the bonding connection and the protective device, respectively.

The safety $S_B(t)$ offered by the protective measure basic insulation is given by Equation (1.7):

$$S_B(t) = e^{-\lambda_B t} \tag{1.7}$$

where λ_B is the failure rate of the basic insulation.

The safety $S_I(t)$ offered by Class I equipment is given by Equation (1.8) (based on Equation (1.4)).

$$\begin{aligned} S_I(t) &= 1 - [(1 - S_B(t)) \cdot (1 - S_1(t)] \\ &= e^{-\lambda_B t} + e^{-\lambda_1 t} - e^{-(\lambda_B + \lambda_1)t}. \end{aligned} \tag{1.8}$$

Equation (1.8) mathematically shows that $S_I(t) > S_B(t)$. However, the same relationship may not hold true for their related risks $r_I(t)$ and $r_B(t)$. The risks in fact depend on $k(t)$ (i.e., the probability that a person touches an energized part), which, as earlier explained, increases with the presence of the metal enclosure for Class I equipment.

1.6 Class II Equipment

Class II equipment refers to electrical devices that ensure both basic and fault protection using double insulation or reinforced insulation. Double insulation includes both basic insulation and supplementary insulation, each applied independently (Figure 1.3).

Reinforced insulation, which can comprise several layers that cannot be tested individually, offers protection against electric shock equivalent to that provided by double insulation. This protective measure prevents the occurrence of dangerous potentials on the equipment upon failure of the basic insulation.

If λ_S is the failure rate of the supplementary insulation, the safety $S_{II}(t)$ against indirect contact offered by Class II equipment is given by Equation (1.9).

Figure 1.3: Double insulated equipment.

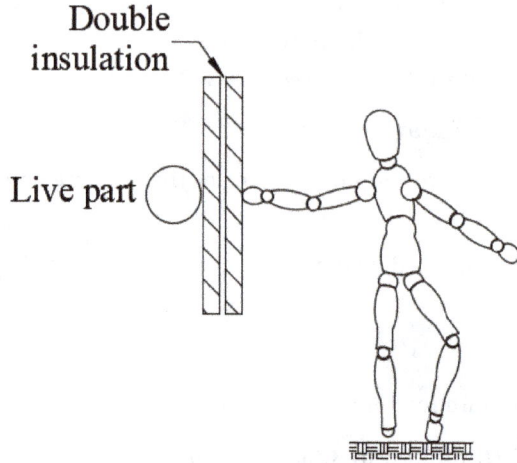

This is a parallel configuration; thus, Equation (1.4) applies.

$$S_{II}(t) = 1 - [(1 - S_B(t)] \cdot [1 - S_S(t)]$$
$$= e^{-\lambda_B t} + e^{-\lambda_S t} - e^{-(\lambda_B + \lambda_S)t}. \tag{1.9}$$

Assuming that $\lambda_1 > \lambda_S$, where $\lambda_1 = \lambda_{EQB} + \lambda_{PD}$, a comparison of Equations (1.8) and (1.9) reveals that $S_{II}(t) > S_I(t)$. Additionally, it can be concluded that $r_{II}(t) < r_B(t)$, where $r_B(t)$ represents the electrical risk associated with basic insulation.

1.7 To Bond or not to Bond

As discussed, the automatic disconnection of supply, the most commonly used protective measure, requires bonding connections of equipment to ensure the operation of protective devices in the event of a ground-fault. Additionally, equipotential bonding is essential in minimizing the potential difference between exposed-conductive parts and extraneous-conductive parts (EXCPs), which are conductive elements that are not part of the electrical installation but are likely to introduce a dangerous potential or the zero potential into the premises. However, not all conductive parts in an electrical system need to be bonded for safety. Equipotential bonding connections should only be employed within electrical equipment, between ECPs, between electrostatically charged objects, between metal objects for lightning protection, and between electrical equipment and metalwork.

In the next sections, we will discuss criteria to determine whether conductive "dead" objects and enclosures of electrical equipment should be bonded or not.

1.7.1 Class I equipment in contact with conductive supports

According to Standard IEC 60050-442[3], a conductive part that can only become live through contact with a faulty exposed-conductive part is not classified an exposed-conductive part itself and, therefore, does not require bonding. For an illustration of this concept, refer to Figure 1.4, showing a Class I equipment supported by a metal bracket.

In the event of a failure in the basic insulation, the ECP becomes energized. However, the metal bracket does not require bonding. If the bracket is already in good contact with the bonded enclosure, no additional bonding connection is necessary. Conversely, if it is not in good contact with the ECP, bonding the bracket would unnecessarily transfer the potential, posing a safety risk to individuals who might come into contact with it.

1.7.2 Extraneous-conductive parts

Extraneous-conductive parts, which have a very low resistance-to-ground R_{EXCP}, may include metal pipework for gas, water, and heating, metallic parts of the building structure, non-insulating floors and walls. A person may come into contact with faulty equipment and an EXCP simultaneously (Figure 1.5).

This contact causes the person's resistance-to-ground, made by the floor or shoes, to be shorted by the EXCP, which places the person's hand at zero potential. The body current is, therefore, maximized with great risk of electrocution. The mitigation of the risk is obtained by bonding the extraneous-conductive parts, as close as possible to their point of entry into the building, with an equipotential connection (EQP) to the building's main ground terminal (MGT). This safety requirement eliminates, or reduces, potential differences between ECPs and EXCPs under ground-fault conditions, regardless of the building's system grounding type (i.e., IT, TT, or TN).

---

[3]IEC 60050-442:1998, "International Electrotechnical Vocabulary – Part 442: Electrical accessories."

Figure 1.4: Class I equipment supported by a metal bracket.

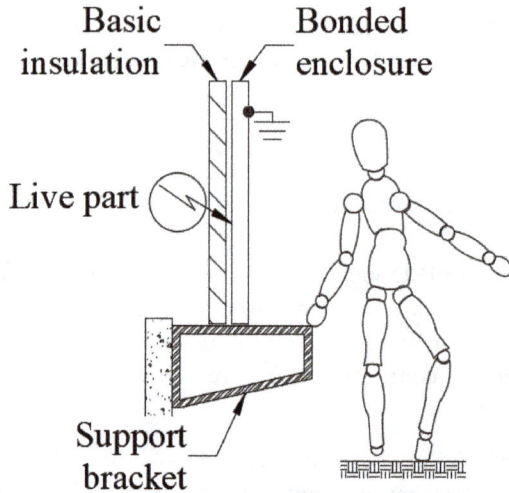

Figure 1.5: Person contacting a faulty equipment and an EXCP.

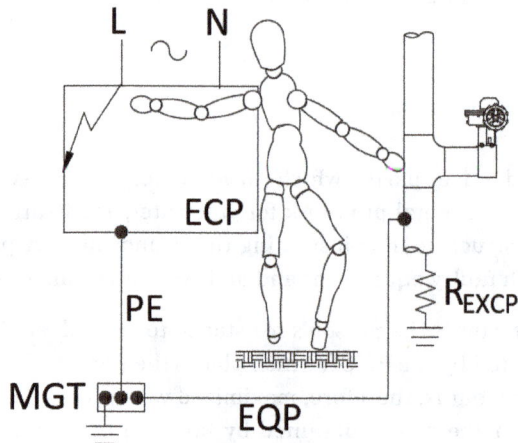

Proper identification of extraneous-conductive parts is essential to ensure equipotentiality during faults. Bonding a mere conductive part, which is not an EXCP, is generally unsafe (Figure 1.6).

Figure 1.6: Unnecessary equipotential bonding of mere metal part.

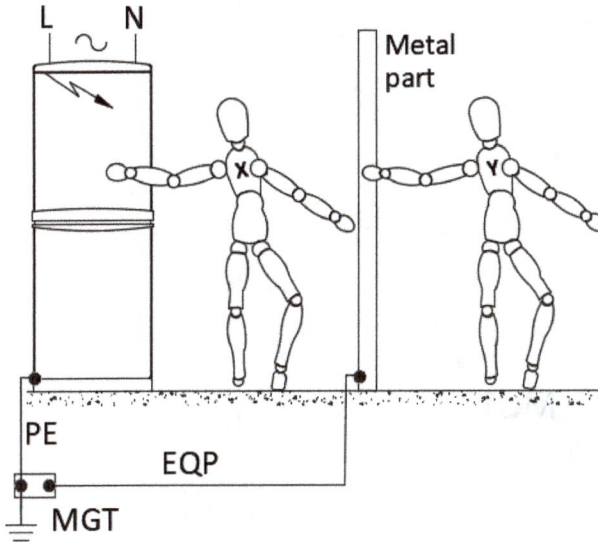

Individual **X** is not at greater risk because of the equipotential connection (EQP) to a metal part (not an EXCP). However, the equipotential connection transfers the fault potential to the metal part energizing it. Consequently, person **Y** is exposed to the risk of electric shock when the appliance fails, a risk that would not otherwise exist.

It may not always be possible to identify an extraneous-conductive part through visual inspection alone. Therefore, measuring the resistance R_{EXCP} between the metal part under investigation and the building ground terminal may be necessary (Figure 1.7).

If the measured resistance is greater than the value calculated using Equation (1.10), the conductive part in question is not considered an extraneous-conductive part, and equipotential bonding is neither required nor necessary for safety.

$$R_{EXCP} = \frac{U_0}{I_B} - R_B. \tag{1.10}$$

U_0 represents the nominal voltage to ground of the installation; R_B and I_B are, respectively, the total body resistance for a hand-to-hand pathway, and the value

Figure 1.7: Resistance test to identify the extraneous-conductive part.

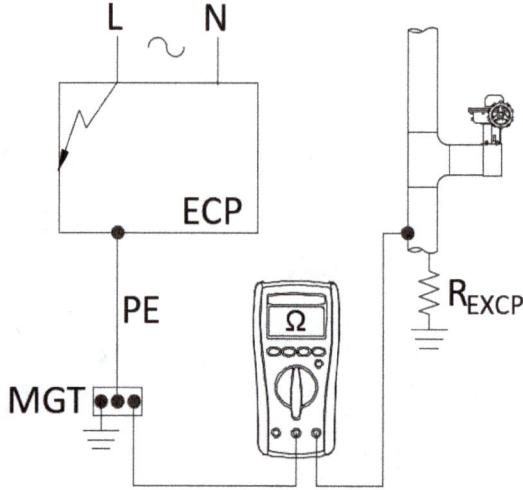

of the body current that should not be exceeded for the safety of the person, both according to IEC 60479-1[4].

The proper value of R_B must be selected by the designer based on several factors, such as the magnitude of the expected touch voltage, the supply frequency, the conditions of the skin (i.e., wet or dry) and the surface area of contact.

The designer must also select the acceptable value of the body current I_B. Typically, 0.5 mA is assumed as the threshold of reaction when touching an energized surface, while the threshold of let-go current is identified at 10 mA. 30 mA is considered the value of current that causes involuntary muscular contractions, difficulty in breathing, reversible disturbances of heart function, and immobilization when withstood for up to 5 seconds, but typically resulting in no organic damage. 30 mA is the most common sensitivity rating of residual current devices (known as GFCIs in the US) for residential applications. As an example, considering $U_0 = 125$ V, $I_B = 10$ mA and $R_B = 1550$ Ω at 50/60 Hz, the value in dry conditions for a large surface area of contact that is not exceeded by 50% of the population, the value of R_{EXCP} is calculated to be 10.9 kΩ.

----

[4]IEC 60479-1:2018: "Effects of current on human beings and livestock."

1.8 Summary

This chapter has emphasized the critical role of designing, implementing, and maintaining robust safety measures in electrical systems to prevent severe incidents, such as electrical shocks and electrocution. Electrical safety must be embedded in the design phase to anticipate and eliminate potential hazards. This proactive approach ensures that protective measures are integrated into electrical systems and equipment, significantly reducing the likelihood of injuries.

A thorough risk assessment is essential in identifying potential hazards and evaluating risks associated with electrical systems. This process aids in implementing safety measures that mitigate these risks effectively, such as insulation and equipotential bonding, which are vital for reducing the risk of electric shock. However, these measures have a non-zero failure rate, necessitating monitoring and maintenance.

Mathematical models help assess the safety and reliability of electrical systems. Understanding and monitoring these metrics enable engineers to improve maintenance strategies and safety protocols. The effectiveness of protective measures in series and parallel configurations determines the overall safety of electrical systems. Series systems depend on the reliability of each component, whereas parallel systems provide redundancy, enhancing safety.

Direct and indirect contacts pose significant risks to individuals. Evaluating the actual electrical risk and selecting appropriate protective measures are essential for ensuring adequate protection against electric shock.

Class I equipment relies on basic insulation and protective bonding, whereas Class II equipment uses double or reinforced insulation for enhanced safety. Both types of equipment require careful consideration of their respective risks and protective measures.

Key Definitions

Basic insulation: The primary layer of insulation provided on live parts to prevent electric shock.

Basic insulation breakdown: Failure of the insulation material, leading to exposure of live parts.

Class I equipment: Electrical equipment that relies on basic insulation and protective bonding to ensure safety.

Class II equipment: Electrical equipment that uses double or reinforced insulation to provide protection against electric shock.

Conductive part: Any part of electrical equipment or systems that can conduct electricity.

Double insulation: An insulation system comprising both basic and supplementary insulation, providing enhanced protection.

Equipotential bonding: The practice of connecting all exposed-conductive-parts and extraneous-conductive-parts to the main grounding terminal to minimize potential differences.

Extraneous-conductive part (EXCP): Conductive elements not part of the electrical installation that could introduce or carry dangerous potentials.

Fault protection: Measures taken to protect against electric shock in the event of an insulation failure.

Hazardous potential: A voltage level that poses a risk of electric shock or injury.

Indirect contact: Contact with an exposed-conductive-part of electrical equipment that has become live due to a fault.

Live part: A part that is energized with voltage and capable of causing an electric shock.

Protective device: A device used to automatically disconnect the power supply in the event of a fault to prevent electric shock.

Protective measures: Methods implemented to prevent electric shock, such as insulation, grounding, and bonding.

Reliability function: A mathematical function representing the probability that equipment will function correctly over time.

Residual current device (RCD): A safety device that disconnects a circuit when it detects an imbalance of electric current, preventing electric shock.

Risk assessment: The process of identifying potential hazards and evaluating the risks associated with electrical systems.

Safety-by-design: The concept of incorporating safety measures into the design phase of electrical systems to anticipate and eliminate hazards.

Series protection system: A system where all protective measures must function correctly simultaneously to ensure safety.

Supplementary insulation: Additional insulation applied independently to provide backup protection in case of failure of the basic insulation.

Touch voltage: The voltage between the conductive part a person is in contact with and the ground.

Voltage limits: The conventional safe voltage thresholds for direct and indirect contact to prevent electric shock.

Bibliography

1. M. Mitolo, "Is it Possible to Calculate Safety: Safety and Risk Analysis of Standard Protective Measures Against Electric Shock," in IEEE Industry Applications Magazine, vol. 15, no. 3, pp. 31-35, May-June 2009.
2. IEC 60364-4-41:2005, "Low-voltage electrical installations - Part 4-41: Protection for safety - Protection against electric shock."
3. IEC 60050-442:1998, "International Electrotechnical Vocabulary - Part 442: Electrical accessories."
4. IEC 60479-1:2018: "Effects of current on human beings and livestock."

CHAPTER

2

Electrical Safety Engineering of Protective Measures not Employing Automatic Disconnection of Supply

---

2.1 Introduction

The IEC 61140[1] technical standard emphasizes that the key to preventing electric shock is to ensure that hazardous live parts are inaccessible and that exposed-conductive parts of equipment, which are not normally live, do not become dangerous in the event of a single fault. A single fault condition occurs when a single protective measure, such as basic insulation, fails. Protecting individuals during such faults is crucial for safety in electrical systems. As previously discussed, this is typically achieved by automatically disconnecting the power supply, which is achieved using bonding conductors (i.e., Class I equipment), which facilitate the operation of protective devices. This active approach reduces the duration of dangerous voltages on a faulty object, thereby mitigating the harmful effects of electric current on the person's body in the case of touch.

This chapter explores *passive* protective measures that maintain the electrical supply even if the basic insulation fails. These measures do not limit the

---

[1]IEC 61140, "Protection against electric shock - Common aspects for installation and equipment."

fault duration but instead use specific wiring and environmental solutions to keep touch voltage at safe levels. They are particularly valuable in scenarios where continuous power is essential.

It is important to note that passive protective measures should only be used in installations managed or supervised by skilled or instructed personnel. A skilled person has the necessary education and experience to recognize and avoid electrical hazards, while an instructed person has been adequately advised or supervised by skilled individuals to perform tasks safely.

2.2 Class II Equipment

An effective method for passive protection consists of adding a second layer of insulation, known as *supplementary insulation*, to the basic insulation of equipment. When combined, these layers are referred to as *double insulation*. Alternatively, a single layer that offers the same level of protection as double insulation can be used, known as *reinforced insulation*. As already anticipated, equipment fitted with double or reinforced insulation are categorized as *Class II*. This protective measure is broadly applicable to various types of equipment, including switchgear and controlgear assemblies, luminaires, appliances, and handheld tools. The two insulating layers are designed to withstand different deteriorating factors, such as temperature and contaminants, ensuring that if one layer fails, the other remains intact, providing continuous protection. Both double and reinforced insulation aim to prevent hazardous voltages from appearing on accessible parts of electrical equipment in the event of a fault in the basic insulation.

Double or reinforced insulation can be used as the sole and exclusive protective measure against electric shock for an installation, regardless of the type of supply or location, but only under effective supervision during normal use. This ensures that no modifications can be made that could compromise its effectiveness.

As earlier discussed, Class II equipment offers enhanced safety, as its failure rate is lower than that of Class I equipment. However, this safety advantage depends on the manufacturer adhering to specified requirements for the insulation material, clearances, and creepage distances relative to the operating voltage. Clearance refers to the shortest distance through the air between conductive parts, while creepage distance is measured along the surface of the insulator. Insulation failure can occur due to material breakdown or electric arcing across its surface under normal operating conditions, especially in polluted or moist environments. The required creepage distances depend on the surface geometry of the insulating parts and the tracking resistance of the insulation material. Properly dimensioned "ribs" can be implemented on the insulation surface to increase the clearance and creepage distance, enhancing the safety of the insulation.

Figure 2.1 presents a graph based on IEC 61439-1[2], illustrating the scenario of Class II switchgear installed either at a construction site or in a clean room environment. At construction sites, insulating materials may be exposed to hazardous conductive pollution (denoted as Pollution Degree 3), while in clean room environments there is no contamination or only dry, non-conductive pollution is present (denoted as Pollution Degree 1). Assume that the insulation material of the equipment has a comparative tracking index (CTI) greater than 600 V (denoted as Material Group I) or between 175 V and 400 V (denoted as Material Group IIIa). The CTI measures a material's resistance to forming conductive tracks when subjected to electrical stress and contamination. Materials with higher CTI values (i.e., Material Group I) perform better and allow for shorter creepage distances.

Figure 2.1: Creepage distance versus rated insulation voltage and pollution degree.

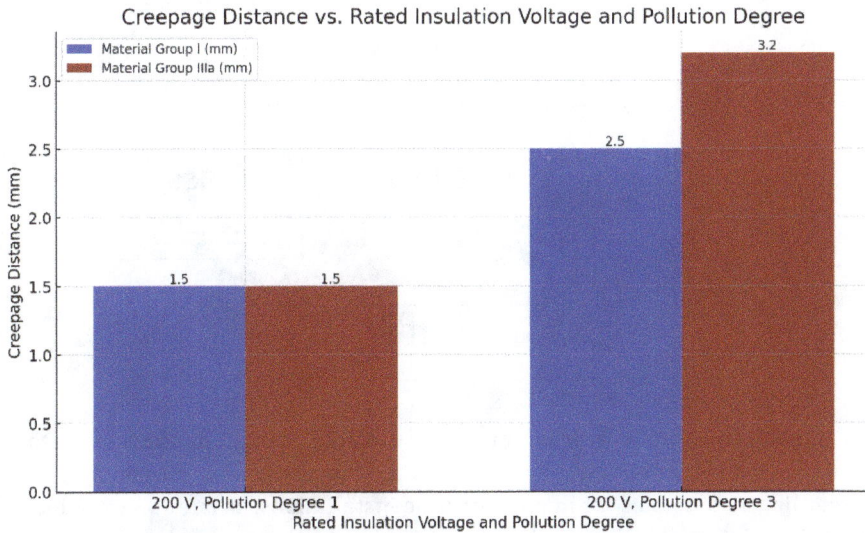

geometry of the insulating parts and the tracking resistance of the insulation

[2]IEC 61439-1:2020, "Low-voltage switchgear and controlgear assemblies - Part 1: General rules."

Figure 2.1 illustrates the critical relationship between the pollution degree of the environment and the required creepage distances based on the properties of insulating materials. This relationship is essential for specifying Class II equipment to ensure reliable and safe operation. The creepage distance requirement increases with the pollution degree for both material groups, reflecting the need for better insulation in more polluted environments. Material Group IIIa requires a larger creepage distance at higher pollution degrees. At a construction site, the required creepage distance within the switchgear must be at least 3.2 mm, which can be reduced to 2.5 mm for better performing insulation.

It is important to note that an installation utilizing double or reinforced insulation as the protective measure (under effective supervision) does not require a protective conductor, referred to as PE (or EGC in the US), to be connected to exposed-conductive parts of Class II equipment. In fact, there is a risk that the protective conductor could energize the enclosure of properly functioning Class II equipment due to the transferred fault potential V_{PE}, as illustrated in Figure 2.2.

Figure 2.2: The protective conductor (PE) transfers the fault potential V_{PE} over the Class II equipment.

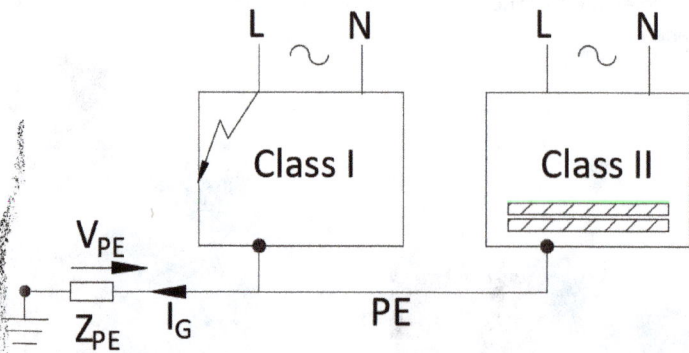

Class II equipment could be connected to a dedicated grounding system that is not shared with other electrical systems. This practice is often used in Class II street lighting systems (or in photo-voltaic installations, as later on examined). Such bonding can facilitate the disconnection of the supply in the event of a failure of double insulation (e.g., a car crashing into a pole), which might otherwise go undetected, exposing pedestrians to the risk of electric shock.

Wiring systems, including cables, are rated as Class II if they have a rated voltage that is at least equal to the nominal voltage of the system and no less than 300/500 V. To provide adequate mechanical protection, Class II wiring systems should be enclosed by either a non-metallic sheath or a non-metallic wireway.

2.3 Electrical Separation

Electrical separation is a protective measure that uses transformers with basic insulation between windings and a unity turn-ratio to galvanically isolate individuals from ground and other circuits. According to the standard IEC 60364-4-41[3] the voltage of the separated circuit must not exceed 500 V. In the event of a ground-fault involving equipment in the separated circuit, a capacitive current may still flow through a person touching the faulted ECP (Fig. 2.3).

Figure 2.3: Capacitive current may flow through a person touching the faulted equipment.

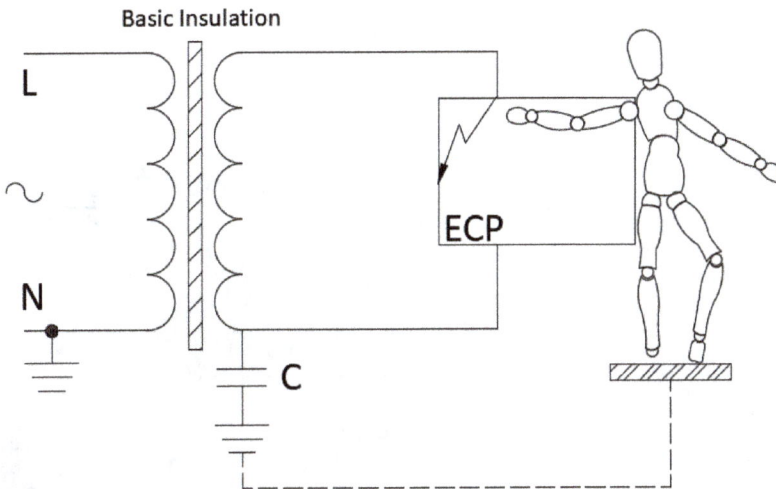

The magnitude of this current depends on the distributed capacitance C of the separated circuit, which depends on its length. According to the standard, to keep this capacitive current at harmless levels, the product of the operating voltage of the separated circuit, in volts, and its length, in meters, must not exceed 100,000 V m, with the length of the circuit not exceeding 500 m. For example, for an operating voltage of 230 V, the maximum safe length of the separated circuit is 434 m.

---

[3]IEC 60364-4-41:2005, "Low-voltage electrical installations - Part 4-41: Protection for safety - Protection against electric shock."

Live parts of the separated circuit must not be connected to ground or other circuits, as this would compromise the electrical separation and expose individuals to the risk of electric shock. In addition, the exposed-conductive parts of the separated circuit, typically Class I equipment, must not be connected to the protective conductor. This is to prevent any possibility of contact with equipment enclosures from other circuits. Such contact would mean that the protection against electric shock would no longer rely solely on the electrical separation, but also on the protective measures applied to the exposed-conductive parts of the other circuits.

The electrical separation measure should only be used to supply one piece of equipment from a single transformer. If multiple pieces of equipment are supplied from the same source, individuals can be exposed to a hazardous potential difference if they simultaneously come into contact with two faulty exposed-conductive parts (Figure 2.4).

Figure 2.4: Double ground-fault causes a hazard when a separation transformer supplies more than one piece of equipment.

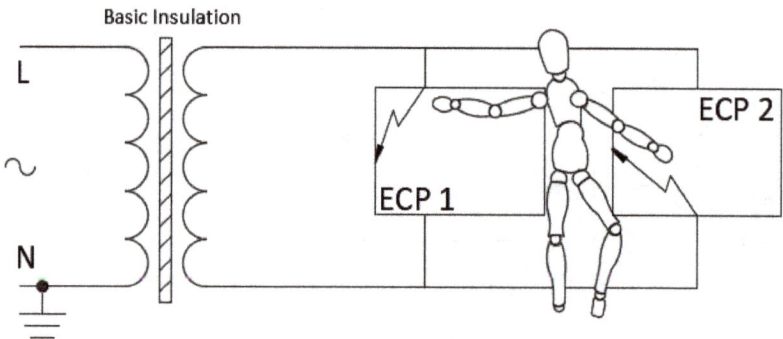

In this scenario, the exposed-conductive parts must be interconnected using insulated, non-grounded equipotential bonding conductors. This equipotential connection neutralizes potential differences, allowing short-circuit currents to flow, which can then be detected and interrupted by overcurrent protective devices.

In areas where the risk of electric shock is higher, the circuit must be isolated using a transformer with double or reinforced insulation between the primary and secondary sides, in accordance with the IEC 61558-2-4 standard[4]. Critical locations where the Class II transformer is required include restrictive

---

[4]IEC 61558-2-4, "Safety of transformers, reactors, power supply units and combinations thereof - Part 2-4: Particular requirements and tests for isolating

conductive locations (discussed in detail in a later chapter), caravan parks and caravans, medical offices, mobile or transportable units, marinas and pleasure craft, and bathrooms.

2.4 Non-conducting Locations

A non-conducting location is a protective measure achieved through the high-impedance nature of the location itself. Insulating walls, floors, and any metalwork entering the location (i.e., EXCPs), significantly limit the circulation of current through a person's body in case of contact with faulty equipment. This measure relies on two layers of protection: the basic insulation of the equipment and the insulation of the location (Figure 2.5).

Figure 2.5: Basic insulation of equipment and insulation of the location as protective measures.

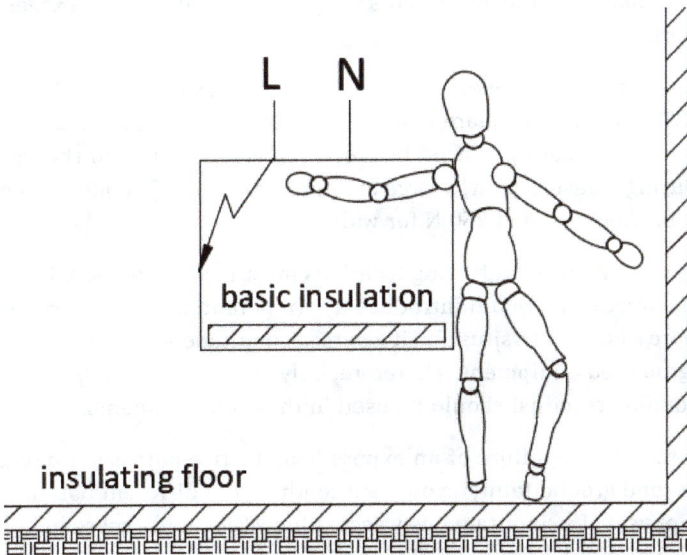

To ensure safety, the resistance-to-ground of insulating floors and walls at every point of measurement, under the conditions specified in IEC 60364-6[5], must not be less than 50 kΩ if the nominal voltage of the installation does

transformers and power supply units incorporating isolating transformers for general applications."

[5]IEC 60364-6, "Low voltage electrical installations - Part 6: Verification."

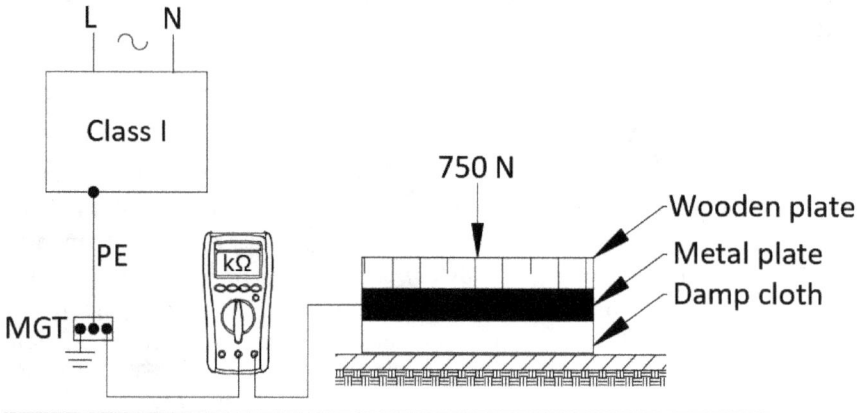

Figure 2.6: Floor resistance-to-ground test.

not exceed 500 V, and at least 100 kΩ if the nominal voltage exceeds 500 V (Figure 2.6).

The electrode for the measurement consists of a square metallic plate with sides of 250 mm, and a square damp cloth, measuring approximately 270 mm on each side. The cloth is placed between the metal plate and the floor being tested. During measurement, a force of approximately 750 N must be applied to the plate for floors, and of 250 N for walls.

Equipment in non-conducting locations must not be connected to a protective conductor, as this would introduce a zero potential into the premises. This could create a hazardous situation if a person simultaneously contacts a live part and the grounded equipment. Therefore, only equipment with basic insulation and no bonding terminal should be used in these environments.

Even though the failure of an exposed-conductive part is not dangerous in itself, a second ground-fault, involving a conductor at different potential (e.g., a different phase) of a simultaneously accessible ECP, could expose individuals to dangerous potential differences (Figure 2.7). An equipotential bonding connection between equipment cannot be carried out, since the ECPs have no bonding terminal. Thus, all ECPs and EXCPs must be secured in specific locations and separated from each other by a spacing of at least 2.5 m. If the proper spacing is unattainable, the interposition of insulating obstacles can be alternatively employed.

The effectiveness of this protective measure against electric shock can be compromised if two faults occur, even if not simultaneously: the failure of the basic insulation of the equipment and the failure of the location's insulation.

Figure 2.7: Equipment in non-conducting locations secured in specific places and separated from each other.

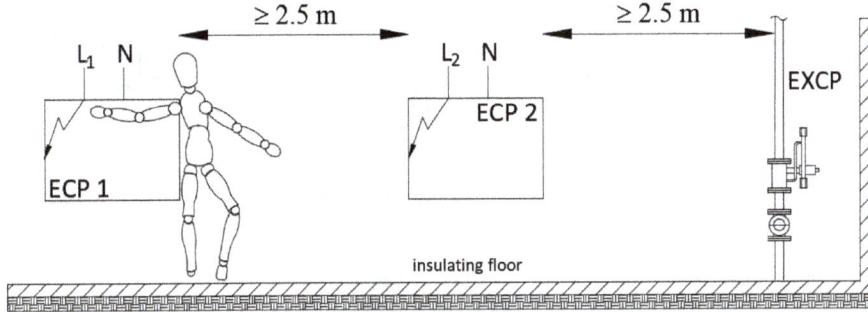

The likelihood of these two faults occurring is based on the failure rates of the insulations, as discussed in Chapter 1.

Due to its delicate nature, protection by a non-conducting location is suitable only in installations strictly supervised by skilled personnel. Key feature of this protection is the absence of any ground reference, which could be inadvertently introduced into the premises by unaware individuals, for example via Class I equipment supplied by extension cords from outside the location.

2.5 Ground-free Local Equipotential Bonding

In ground-free local equipotential bonding locations, the floor must be insulated from the ground, similar to non-conducting locations. Dangerous touch voltages are prevented by using local equipotential bonding conductors (EBC). These bonding conductors connect simultaneously accessible equipment, which must be of Class I type (Figure 2.8).

In these locations, the supply is not automatically disconnected under ground-fault conditions, therefore faulty equipment remains energized. For this reason, the equipotential bonding conductor must not be connected to the main grounding system of the building, as this would transfer faults from the ground-free area to grounded equipment elsewhere.

During a ground-fault, a person entering the ground-free location from a grounded area may be exposed to touch or step voltages. This risk can be mitigated by installing an insulating mat at the access points to the ground-free location.

Figure 2.8: Ground-free local equipotential bonding location.

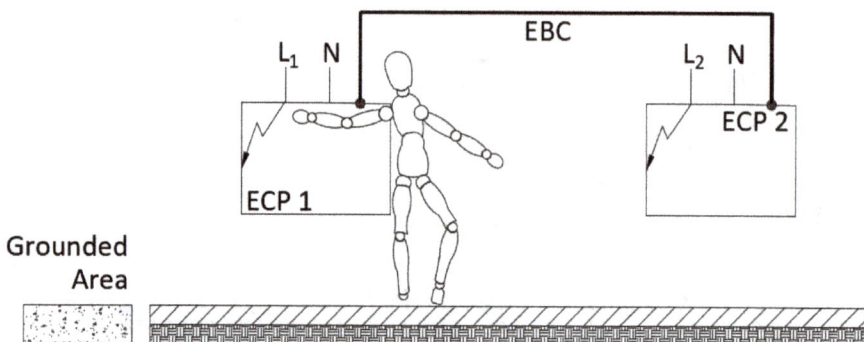

The safety of this protective measure is compromised if two faults occur: the failure of the basic insulation of the equipment and the failure of the equipotential bonding connection. Thus, regular inspections and maintenance of the equipotential bonding system are crucial to ensure all connections remain intact. Clear warning signs should be installed at the entrances of ground-free locations to alert personnel of potential risks and necessary precautions. Additionally, it is important to provide training for workers on the specific safety protocols and emergency procedures related to this protective measure.

2.6 Extra-low Voltage Systems

Protection by extra-low voltage (ELV) utilizes non-hazardous supply voltages to ensure both basic and fault protection. ELV systems are restricted to maximum voltages of 50 V a.c. (r.m.s.) and 120 V d.c., measured between line and ground and between lines. This measure is broadly applicable but is especially recommended for high-risk environments, such as restrictive conductive locations (discussed in detail in a dedicated chapter). As is further discussed in the next chapter, there are fundamentally two types of extra-low voltage systems designed to ensure protection against contact hazards: safety extra-low voltage (SELV) and protective extra-low voltage (PELV). SELV systems employ galvanic isolation and have no intentional connections to ground of the source or the exposed-conductive-parts. PELV systems, while grounded, adhere to stringent requirements for insulation, equipotential bonding, and protective separation to mitigate risks associated with transferred potentials during ground faults.

A third type, functional extra-low voltage (FELV), is primarily functional and does not ensure protection against both direct and indirect contact hazards.

ELV circuits must be effectively insulated from higher voltage circuits, which requires the use of Class II transformers. It is important to note that relying on the sole basic insulation for the transformer would pose electric shock risks to individuals. As shown in Figure 2.9, if the basic insulation at the transformer fails (fault denoted as 1), it can also compromise the insulation of the SELV equipment (fault denoted with 2) (e.g., lighting systems in swimming pools), which typically lack the dielectric strength for higher voltage. This failure results in the operating voltage V_{LN} being transferred to users due to a single failure.

Figure 2.9: Basic insulation failure in a Class I transformer.

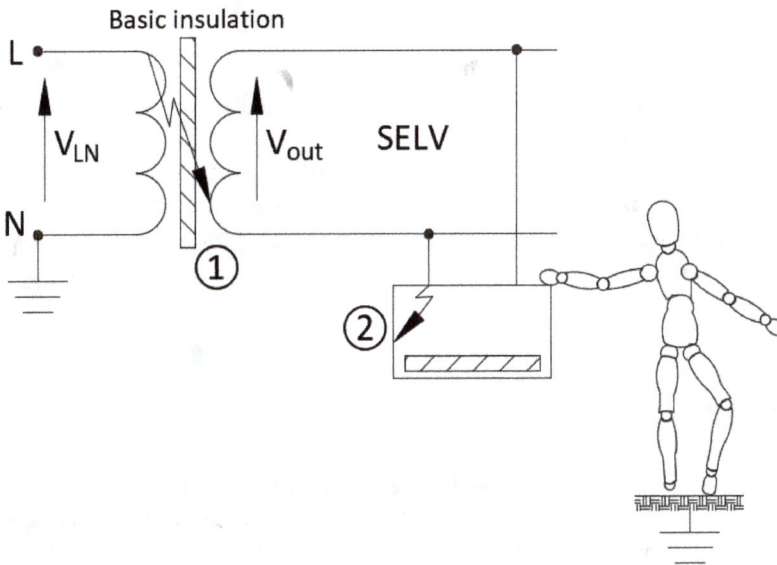

SELV sources are not permitted to be connected to ground. The reason behind this prohibition is exemplified in Figure 2.10.

In normal, dry, operating conditions, a person in contact with either secondary pole of the SELV circuit is not in danger, the system being isolated for ground. However, if either secondary pole is accidentally grounded, for example due to a fault, the person is exposed to the voltage V_{out}. In wet conditions, this voltage exposure may be problematic, even for ELV voltages.

Figure 2.10: Secondary pole ground-fault in SELV systems.

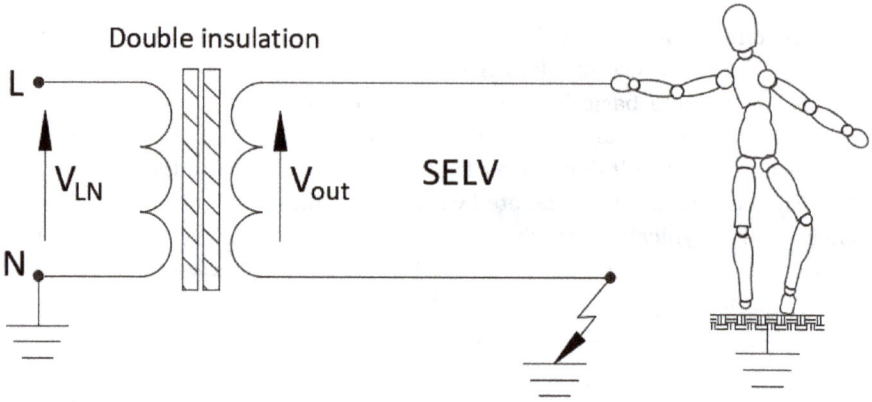

In PELV systems, where supplies are intentionally grounded to the same grounding electrode as non-PELV circuits, a person in contact with a secondary pole is always exposed to the voltage V_{out}. Additionally, it will be discussed later that ground faults involving non-PELV equipment may expose individuals to a touch voltage $V_{Tp} = V_{out} + V_{PE}$, where V_{PE} is the voltage drop on the protective conductor resulting from the fault, which poses an increased risk of electric shock.

2.7 Summary

Passive protective measures provide safety by reducing the risk of electric shock while maintaining power continuity. These measures are particularly advantageous in environments where continuous power is critical, such as medical facilities, data centers, and industrial operations. However, the effectiveness of these passive measures relies heavily on meticulous design, strict adherence to safety guidelines, and the supervision of skilled personnel.

Non-conducting locations leverage high-impedance environments to prevent current circulation through a person's body. These locations require insulating floors and walls with high resistance-to-ground values. However, this measure's effectiveness can be compromised if basic insulation fails or if grounded parts are inadvertently introduced, underscoring the need for ongoing supervision and control.

Ground-free local equipotential bonding involves connecting simultaneously accessible Class I equipment through equipotential bonding conductors, creating a Faraday cage effect. This method prevents potential differences between metal parts but requires careful management to avoid transferring fault potentials from the ground-free area to grounded equipment elsewhere in the building. One challenge lies in ensuring that ground-faults do not expose individuals to touch or step voltages at the interfaces with standard installations.

Double or reinforced insulation provides continuous protection even if one layer fails. This approach is robust in preventing hazardous voltages from appearing on accessible parts of electrical equipment. However, regular maintenance is essential to preserve the integrity of the insulation, and unauthorized modifications must be avoided to maintain safety.

ELV systems are particularly recommended for high-risk environments. Effective insulation from higher voltage circuits, use of safety transformers, and wires with double or reinforced insulation are critical for ELV circuits.

The implementation of passive protective measures demands a thorough understanding of electrical safety principles and the potential hazards associated with each environment. Regular inspections and maintenance are critical to ensure all safety measures remain intact and effective. Additionally, clear warning signs and training for personnel on specific safety protocols and emergency procedures are vital to mitigate risks when the automatic disconnection of supply is not present.

Key terms

Class I equipment: Electrical equipment that relies on basic insulation and a protective earthing connection to prevent electric shock in case of a fault.

Class II equipment: Electrical equipment with double or reinforced insulation that does not require a protective grounding connection, ensuring safety even if one insulation layer fails.

Double insulation: The use of two layers of insulation (basic and supplementary) to protect against electric shock, commonly found in Class II equipment.

Electrical safety: Measures and practices designed to prevent electric shock, fire, and other hazards associated with the use of electrical systems and equipment.

Electrical separation: A protective measure using transformers to isolate individuals from ground and other circuits, preventing electric shock by maintaining galvanic isolation.

Electric shock: The physiological effect of electric current passing through a person's body, which can cause injury or death.

Equipotential bonding: The practice of connecting conductive parts to a common bonding network to eliminate potential differences and reduce the risk of electric shock.

Extra-low voltage (ELV): A low voltage system that reduces the risk of electric shock by using supply voltages not exceeding 50 V a.c. or 120 V d.c.

Fault duration: The time period during which a fault condition exists in an electrical system, with longer durations posing greater risks.

Ground-free local equipotential bonding: A protective measure connecting Class I equipment within a specific area to prevent potential differences, without linking to the main grounding system.

Non-conducting locations: Areas designed with high-impedance surfaces to prevent current circulation through a person's body, enhancing electrical safety.

Passive protective measures: Safety measures that do not automatically disconnect the power supply during a fault but instead use specific designs to keep touch voltage at safe levels.

Pollution degree: A classification of the environmental contamination level, affecting the insulation requirements for electrical equipment.

Protective conductor: A conductor used to connect the exposed-conductive parts of equipment to the main grounding system, reducing the risk of electric shock.

Reinforced insulation: A single layer of insulation that provides equivalent protection to double insulation, ensuring safety in Class II equipment.

Safety: The state of being protected from electrical hazards, including electric shock, fire, and other dangers associated with electrical systems.

Bibliography

1. IEC 61140:2016, "Protection against electric shock - Common aspects for installation and equipment."
2. IEC 60050-195-01-23, "Earthing and protection against electric shock."

3. G. Parise, "A New Summary on the IEC Protection Against Electric Shock," *in IEEE Transactions on Industry Applications*, vol. 49, no. 2, pp. 1004-1011, March-April 2013.

4. IEC 60364-4-41:2005, "Low-voltage electrical installations - Part 4-41: Protection for safety - Protection against electric shock."

5. M. Mitolo, "Is it Possible to Calculate Safety: Safety and Risk Analysis of Standard Protective Measures Against Electric Shock," *in IEEE Industry Applications Magazine*, vol. 15, no. 3, pp. 31-35, May-June 2009.

6. IEC 61439-1:2020, Low-voltage switchgear and controlgear assemblies - Part 1: General rules."

7. G. Parise, L. Martirano and M. Mitolo, "Electrical Safety of Street Light Systems," *in IEEE Transactions on Power Delivery*, vol. 26, no. 3, pp. 1952-1959, July 2011.

8. M. Mitolo, "On Outdoor Lighting Installations Grounding Systems," *in IEEE Transactions on Industry Applications*, vol. 50, no. 1, pp. 33-38, Jan.-Feb. 2014.

9. CEI 64-8:2019, "Low voltage electrical installations."

10. IEC 61558-2-4:2021, "Safety of transformers, reactors, power supply units and combinations thereof - Part 2-4: Particular requirements and tests for isolating transformers and power supply units incorporating isolating transformers for general applications."

11. M. Mitolo and P. Montazemi, "Electrical Safety in the Industrial Workplace: An IEC Point of View," in IEEE Transactions on Industry Applications, vol. 50, no. 6, pp. 4329-4335, Nov.-Dec. 2014.

3

Electrical Safety Engineering of PELV, SELV and FELV Systems

3.1 Introduction

As discussed in the previous chapter, the protection by extra-low voltage (ELV), achieved through the supply of electrical systems with non-hazardous secondary voltages, is a critical measure for safeguarding against both direct and indirect contact. This approach is generally applicable, but it is particularly recommended for environments with increased risk, such as wet locations or restrictive conductive locations. Generally, extra-low voltages are limited to maximum values of 50 V a.c. (r.m.s.) and 120 V d.c.; however, in special locations, the IEC 60364-7 series standards impose more stringent limitations on their values. For example, IEC 60364-7-702[1] limits the system nominal voltage to 12 V a.c. or 30 V d.c. in swimming pools.

---

[1]IEC 60364-7-702:2010, "Low-voltage electrical installations – Part 7-702: Requirements for special installations or locations – Swimming Pools and Fountains."

Approved sources for ELV systems include a safety isolating transformer (equipped with double or reinforced insulation between the primary and secondary sides) and a motor generator. An electrochemical source, like a battery, or any other source that operates independently of a higher voltage circuit may also be used.

Electrical equipment operating at supply voltages restricted to extra-low values fall under the classification of Class III equipment. In the event that the basic insulation between transformer windings becomes compromised, such as due to a fault or a short circuit, this results in the subsequent breakdown of insulation within the Class III appliance. Typically, these devices lack the dielectric strength necessary for higher voltages. Consequently, as previously discussed, ELV sources require double or reinforced insulation to more effectively safeguard extra-low voltage circuits against inadvertent contact with other circuits that may carry higher voltages.

3.2 Safety Extra-low Voltage (SELV) systems

According to IEC 60364-4-41[2], safety extra-low voltage (SELV) systems must meet specific requirements for both basic and fault protection. These requirements include ensuring basic insulation between live parts and other ELV circuits, as well as between live parts and the ground. Additionally, there must be double or reinforced insulation of wiring systems of SELV (and PELV) circuits from circuits not being at extra-low voltage when their physical separation cannot be arranged.

Furthermore, circuits and exposed-conductive parts of equipment must not be connected to ground. If ECPs of SELV circuits come into contact with ECPs of other circuits, the protection against electric shock would no longer rely solely on SELV protection but also on the protective measures applicable to the other equipment.

For safety isolating transformers, protective separation can also be achieved using basic insulation combined with a grounded metallic screen between the primary and secondary windings. This solution, although allowed, is less safe than double insulation. If the insulation between the secondary winding and the metal screen fails (fault denote with 1 in Figure 3.1), the secondary winding would become grounded.

---

[2]IEC 60364-4-41, "Low-voltage electrical installations - Part 4-41: Protection for safety - Protection against electric shock."

Figure 3.1: Failure of the insulation between the secondary winding and the grounded metal screen.

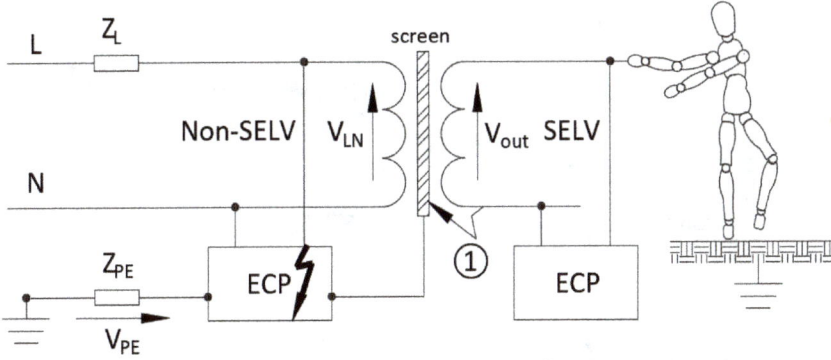

In this situation, a ground-fault occurring on any non-SELV exposed-conductive part would pose a threat to the SELV system. In the worst-case scenario, the entire safety transformer output voltage V_{out} would add to the voltage drop V_{PE} across the protective conductor (PE), such as $V_{Tp} = V_{out} + V_{PE}$. This undermines the SELV system's purpose of maintaining harmless voltages. This over-voltage would persist until the protective device of the higher voltage circuit clears the fault. Consequently, in this scenario, protection against electric shock within the SELV system would become contingent upon the protective provisions of the non-SELV system.

In the scenario depicted in Figure 3.1, V_{PE} can be calculated using Equation (3.1).

$$V_{PE} = V_{LN} \frac{Z_{PE}}{Z_L + Z_{PE}} \tag{3.1}$$

where V_{LN} is the nominal line-to-ground voltage, Z_{PE} is the impedance of the protective conductor (PE) (also referred to as equipment grounding conductor in the US NFPA 70[3]), between the point of the fault and the source, Z_L is the impedance of the line conductor up to the point of the fault. In this case, V_{PE} will persist until the protective device of the non-SELV circuit clears the fault condition. Equation (3.1) underscores the critical role played by the impedances Z_{PE} and Z_L in determining the magnitude of the hazardous potential transferred to the SELV system during a ground fault scenario.

For safety reasons, standard IEC 60364-4-41 explicitly prohibits the inter-connection of ECPs in the SELV circuit with the grounding system, protective

---

[3]NFPA 70 National Electrical Code.

conductors or any equipment of non-SELV circuits. This interconnection would allow the transfer of hazardous potentials V_{PE} originating from other locations, which can be calculated with Equation (3.1), thereby decreasing the overall system safety (Figure 3.2).

Figure 3.2: Unsafe interconnection of ECP in SELV circuit with ECP of the non-SELV circuit.

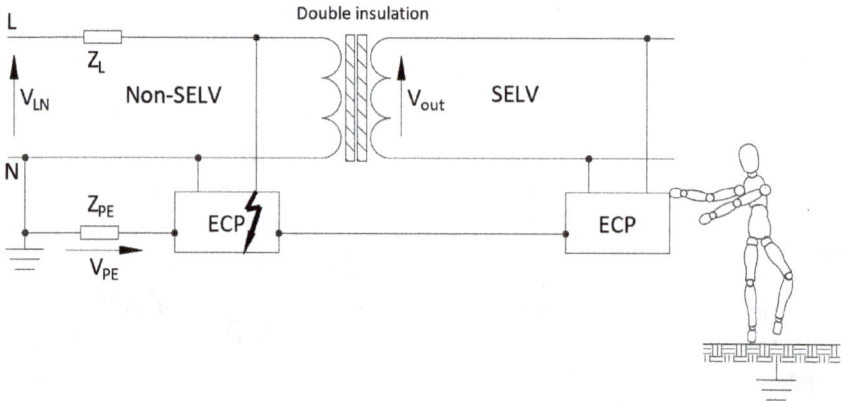

To ensure complete electrical separation and prevent inadvertent inter-connections, plugs and receptacles employed in SELV circuits must be devoid of grounding conductor contacts. Plugs must be mechanically incompatible with receptacles designated for non-SELV systems, preventing their insertion. Conversely, receptacles in SELV systems must be constructed in a manner that precludes the admission of plugs associated with other voltage systems.

3.2.1 Safety considerations and voltage constraints in SELV systems

A key factor enabling the enhanced safety of SELV systems is the galvanic separation between the primary and secondary windings of the safety isolating transformer. This separation results in the absence of any ground reference on the secondary side. Consequently, in the event of direct contact with a live terminal, no significant current can flow through a person's body, as there is no ground return path established.

Extra-low voltage systems operate under the fundamental premise that their voltage levels are typically too low to pose an electric shock hazard under normal dry conditions, when the body resistance of an exposed person remains within normal physiological limits. This constraint effectively minimizes the

Figure 3.3: Transferred potential due to non-PELV equipment ground-fault.

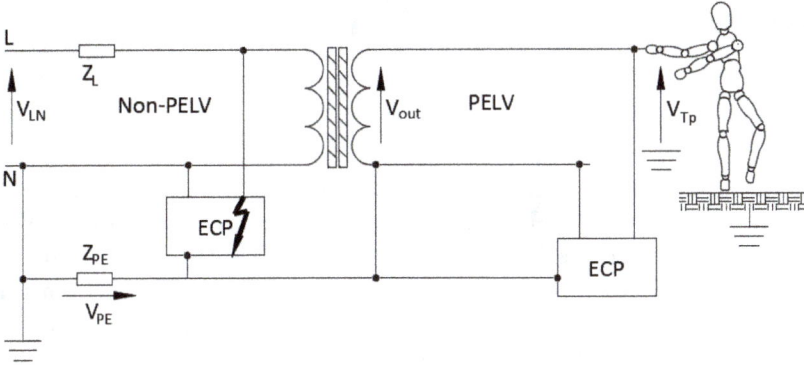

risk of harmful electric currents passing through the human body, even if there is simultaneous direct contact with both transformer terminals.

When the nominal voltage of a SELV system does not exceed 25 V a.c. or 60 V d.c., basic insulation of live parts is not deemed necessary for protection against direct contact. It is crucial to note that, in wet conditions, the inherent safety advantages of extra-low voltages may be compromised. Specifically, in the event of a fault that grounds one of the transformer's output terminals, direct contact with the other live terminal could potentially result in a hazardous electric shock scenario, as exemplified in Figure 2.10.

IEC 60364-4-41 recognizes that if the nominal voltage of the SELV system does not exceed 12 V a.c. or 30 V d.c., the need for basic protection, such as insulation of live parts against direct contact, is deemed unnecessary, even in wet locations.

3.3 Protective Extra-low-voltage Systems (PELV)

Protective extra-low voltage (PELV) refers to an extra-low voltage electrical circuit that is grounded while still meeting all the requirements for SELV. The exposed-conductive-parts of equipment supplied by the PELV circuits are also grounded. PELV systems are suitable for circuits that require grounding, for instance, for functional reasons or for safety purposes, as is the case with control circuits. However, it's important to note that the grounded implementation of PELV systems introduces the risk of transferred potentials during ground-fault scenarios in non-PELV systems (as depicted in Figure 3.3).

In these ground-fault situations, the terminal potential V_{Tp} of the safety transformer will exceed the nominal PELV voltage, as determined by Equation (3.2):

$$V_{Tp} = V_{PE} + V_{out} = V_{LN} \frac{Z_{PE}}{Z_L + Z_{PE}} + V_{out} \tag{3.2}$$

where V_{out} is the nominal output voltage of the PELV system and V_{PE} is the voltage drop across the protective conductor. This over-voltage condition at the PELV terminal, persisting until the protective device of the non-PELV circuit clears the fault, poses an increased risk of electric shock if direct contact is made with the live PELV terminal during such events. The severity of this risk is influenced by two key factors: the failure rate of Class I equipment connected on the primary side of the safety transformer and the possible absence of the basic insulation of PELV, as per the relaxed requirements for extra-low-voltage systems.

The safety improves if the person is not standing in a zero potential region but rather in the equipotential area (Figure 3.4).

To achieve this, all exposed-conductive-parts and extraneous-conductive-parts, such as water pipes, conductive floors, should be interconnected and bonded to a common ground terminal through equipotential bonding. Within this equipotential area, an individual who accidentally contacts a live conductor would only experience the extra-low voltage V_{out} of the PELV system. Essentially, the person's feet are at the potential V_{PE} rather than zero.

Figure 3.4: Person in the equipotential area in PELV systems.

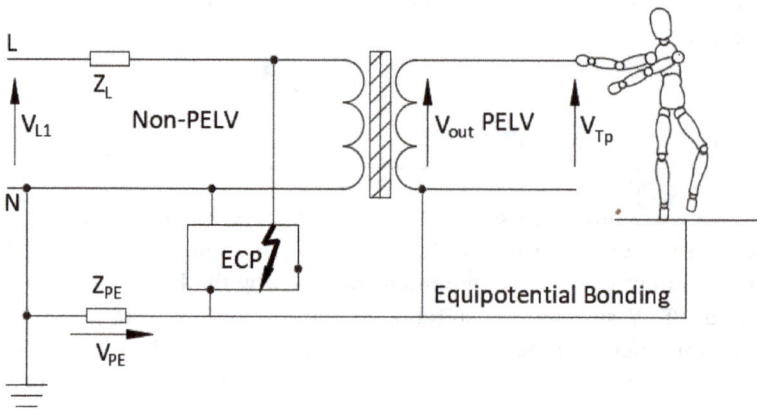

According to IEC 60364-4-41, when equipotential bonding connections are in place, basic insulation of live parts is generally unnecessary under dry conditions if the nominal voltage of the PELV system does not exceed 25 V a.c. or 60 V d.c. In all other cases, basic protection is not required if the nominal voltage of the PELV system does not exceed 12 V a.c. or 30 V d.c.

3.3.1 PELV systems in control circuits for machinery

In the context of control circuits for machinery, an additional critical safety requirement is to prevent unexpected start-ups or unintended shutdowns, even in the presence of one or more ground faults. Such events could pose great hazards to personnel and potentially damage the equipment.

While control circuits should ideally operate at extra-low voltage levels for enhanced safety, the use of SELV circuits may not always be suitable. The absence of a reference ground connection, a key safety feature of SELV systems, can hinder the effective clearing of ground faults through protective devices. This limitation is illustrated in Figure 3.5, where a double ground-fault toward the chassis of the control panel short-circuits the normally open contact M.

As a result, coil M becomes energized regardless of the push button's position (denoted as PB), leading to an unexpected and potentially hazardous motor start-up.

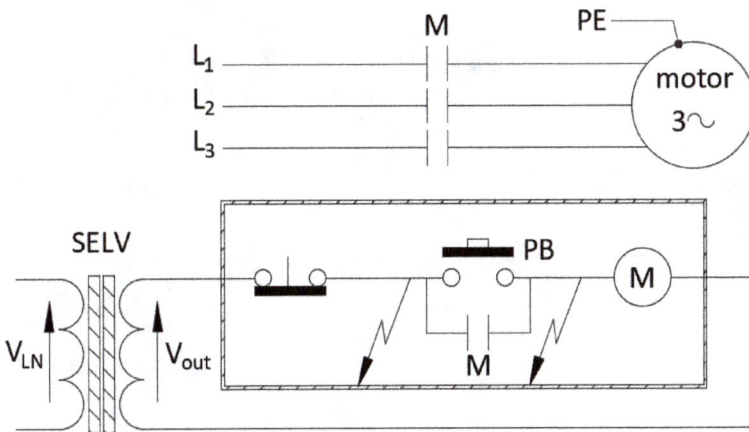

Figure 3.5: Double ground-fault energizing the coil M.

PELV systems are better suited for control circuits. The reason lies in two crucial features of PELV systems: a ground reference point and the ability to bond the control panel's chassis to this ground reference. In this grounded and bonded configuration, the first ground fault effectively creates a short circuit between the live conductor and the grounded chassis. This short circuit ensures the proper operation of protective devices, typically fuses, promptly disconnecting the supply.

3.4 Functional Extra-low-voltage (FELV) Systems

Functional extra-low voltage (FELV) systems are only utilized for functional purposes, such as machine control systems, rather than for safety purposes. FELV systems do not adhere to the fundamental safety requirements mandated for SELV or PELV systems, such as the use of safety isolating transformers and double insulation from higher voltage systems. Consequently, there is a risk associated with the failure of the basic insulation between the windings of the transformer or adjacent higher voltage circuits. In such instances, the higher voltage could transfer to the secondary side of the transformer, increasing dangerously the voltage-to-ground of the circuit (Figure 3.6).

Beyond the direct contact risks, FELV systems are also vulnerable to indirect contact hazards. Specifically, a failure of the basic insulation in non-FELV equipment may cause a voltage transference that can effectively compromise the insulation of FELV equipment (e.g., relays, contactors, PLCs, etc.). In such

Figure 3.6: Failure of the basic insulation between the windings in the FELV system.

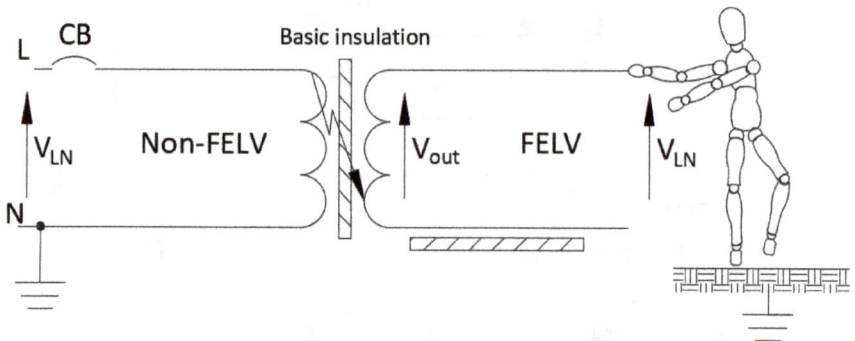

scenarios, individuals in contact with this FELV equipment would be at risk of electric shock from the higher voltage source (Figure 3.7).

Figure 3.7: Ground-fault in the non-FELV circuit propagates and induces a ground-fault in the FELV circuit.

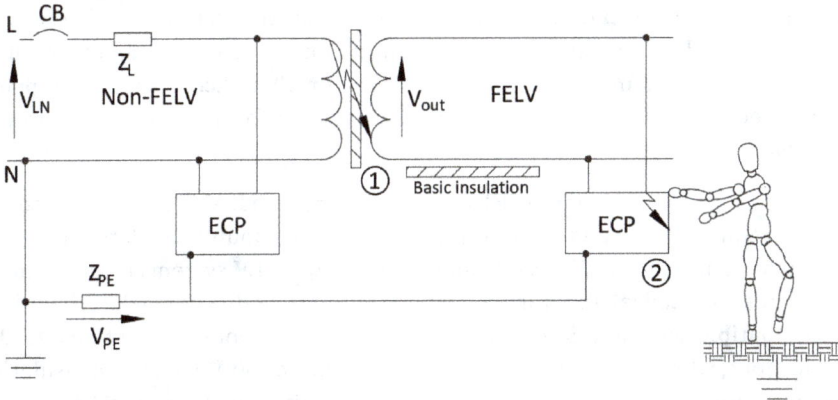

In this scenario, it is crucial that exposed-conductive parts of the FELV circuit must be connected to the protective conductor of the primary circuit of the source, provided that the primary circuit is equipped with automatic disconnection of supply protection. Figure 3.7 illustrates a ground-fault in the non-FELV circuit (denoted as 1) that propagates and induces a subsequent ground-fault in the FELV equipment (denoted as 2). In this situation, the two ground faults are not independent events but are causally related. This cascading effect highlights the susceptibility of FELV systems to insulation failures and voltage transferences originating from interconnected higher voltage circuits.

This grounding arrangement facilitates the proper operation of protective devices in the primary circuit (e.g., circuit breaker, denoted as CB), which can automatically disconnect the supply. This configuration reduces the hazardous voltage exposure to V_{PE}, as shown in Figure 3.7. In the absence of the equipment interconnection to the protective conductor of the source, the prospective touch voltage would be the greater value V_{LN}.

3.5 Summary

Extra-low voltage (ELV) systems are essential for ensuring electrical safety by supplying non-hazardous secondary voltages, protecting against direct and indirect contact. ELV systems are particularly recommended for high-risk environments like wet locations and restrictive conductive locations. The IEC 60364-7 series standards set specific voltage limits for these environments, such as limiting system nominal voltage to 12 V a.c. or 30 V d.c. in swimming pools. Approved ELV sources include safety isolating transformers, motor generators, and electrochemical sources like batteries.

Safety extra-low voltage (SELV) systems adhere to IEC 60364-4-41 standard, which requires basic and fault protection, including double insulation between live parts and other ELV circuits and the ground. SELV systems must maintain reinforced electrical separation, and their plugs and receptacles should be incompatible with non-SELV systems to avoid hazardous connections. In dry conditions, when voltages do not exceed 25 V a.c. or 60 V d.c., basic insulation is unnecessary; in wet conditions the voltage limits are lowered to 12 V a.c. or 30 V d.c.

Protective extra-low voltage (PELV) systems meet SELV requirements but are grounded, making them suitable for control circuits. However, grounding introduces the risk of transferred potentials during ground faults. To mitigate this, equipotential bonding is used to keep potential differences under fault conditions low, ensuring safety. PELV systems are preferred for machinery control circuits to prevent unintended start-ups or shutdowns during ground-faults. Grounding and bonding the chassis ensure that protective devices can effectively clear faults.

Functional extra-low voltage (FELV) systems are used for purposes such as machine control and do not meet the safety requirements of SELV or PELV systems. FELV systems are vulnerable to direct and indirect contact hazards due to the potential transfer of higher voltages. To mitigate risks, the proper grounding of FELV exposed-conductive parts to the primary circuit's protective conductor ensures that protective devices can disconnect the supply, reducing hazardous voltage exposure.

Key terms

Basic insulation: The insulation applied to live parts to provide basic protection against electric shock.

Class III equipment: Electrical devices operating at supply voltages restricted to extra-low values, typically lacking the dielectric strength for higher voltages.

Direct contact: Contact of persons or livestock with live parts that can result in an electric shock.

Double insulation: A safety measure involving two layers of insulation to protect against electric shock.

Equipotential bonding: The practice of connecting all exposed-conductive parts and extraneous-conductive parts to a common ground to maintain the same electrical potential and reduce shock hazards.

Extra-low voltage (ELV): A voltage level not exceeding 50 V a.c. (r.m.s.) or 120 V d.c., used to enhance safety by minimizing the risk of electric shock.

Fault protection: Measures taken to protect against electric shock resulting from a fault condition, such as a ground-fault.

Functional extra-low voltage (FELV): ELV systems used for functional purposes without meeting the stringent safety requirements of SELV or PELV systems.

Ground-fault: An unintended connection between an energized conductor and ground, which can pose a risk of electric shock.

Indirect contact: Contact of persons or livestock with exposed-conductive parts that have become live under fault conditions.

Isolation transformer: A transformer used to transfer electrical power from a source of alternating current (a.c.) to some equipment or device while isolating the powered device from the power source for safety.

Protective conductor (PE): A conductor used to connect the exposed-conductive parts of equipment to the main grounding terminal to ensure safety by providing a path for fault currents.

Protective extra-low voltage (PELV): An extra-low voltage system that is grounded while meeting all the requirements for SELV, often used in control circuits requiring grounding.

Safety extra-low voltage (SELV): An extra-low voltage system designed to ensure safety by preventing electric shock through insulation and separation from higher voltage circuits.

Safety isolating transformer: A transformer providing protective separation (i.e., double insulation) between its primary and secondary windings to ensure that the secondary circuit is safe from higher voltages.

Secondary voltage: The output voltage supplied by an isolation transformer or other ELV sources, typically non-hazardous and used for safety purposes.

Bibliography

1. IEC 61140:2016, "Protection against electric shock - Common aspects for installation and equipment."
2. IEC 60364-4-41:2005, "Low-voltage electrical installations - Part 4-41: Protection for safety - Protection against electric shock."
3. IEC 60364-7-702:2010, "Low-voltage electrical installations – Part 7-702: Requirements for special installations or locations – Swimming Pools and Fountains."
4. IEC
5. 60050 - 195-06-36, "Earthing and protection against electric shock."
6. IEC 61558-2-6:2021, "Safety of transformers, reactors, power supply units and combinations thereof - Part 2-6: Particular requirements and tests for safety isolating transformers and power supply units incorporating safety isolating transformers for general applications."
7. NFPA 70: 2023, "National Electrical Code."
8. M. Mitolo, "Is it Possible to Calculate Safety: Safety and Risk Analysis of Standard Protective Measures Against Electric Shock," in IEEE Industry Applications Magazine, vol. 15, no. 3, pp. 31-35, May-June 2009.
9. IEC 60204-1:2016, "Safety of machinery - Electrical equipment of machines - Part 1: General requirements."

4

Electrical Safety Engineering of Conducting Locations with Restricted Movements

4.1 Introduction

Conducting locations with restricted movements (CLRs), as defined by IEC 60364-7-706[1], are areas predominantly enclosed by extraneous-conductive parts (EXCPs)[2]. As discussed in previous chapters, EXCPs are conductive elements not part of the electrical installation but likely to introduce dangerous potentials or the zero potential (i.e., local ground) into the location. In such environments, it is likely that a person may come into contact with the EXCPs at multiple points on their body, with minimal possibility to disrupt such contact.

Typical examples of CLRs include settings with a significant presence of metallic or conductive materials that are well grounded, such as transmission towers, metal tanks, damp tunnels, and similar locations in good contact with

---

[1]IEC 60364-7-706, "Low-voltage electrical installations - Part 7-706: Requirements for special installations or locations - Conducting locations with restricted movement."

[2]M. Mitolo: "Protecting Electrical Workers in Conducting Locations with Restricted Movements," Distributed Generation & Alternative Energy Journal (DGAEJ). Vol. 39_4, 1–12.

the earth. Extended bodily contact can also occur because of the tasks workers must perform. For instance, a transmission tower qualifies as a CLR because linepersons, during their work, inevitably maintain contact with the metal structure.

The electrical hazard within CRLs arises from electrical equipment, whether fixed or hand-held. If equipment experiences an electrical breakdown, workers face an increased risk of electric shock. This risk is heightened because the body's resistance-to-ground is significantly reduced due to unavoidable contact with the EXCPs. Consequently, the threshold for ventricular fibrillation, a life-threatening heart rhythm disturbance, can be lower in the event of contact with an energized part. Moreover, the typical current pathway from hands to feet for individuals in contact with an energized object may not be the only possible route. A more dangerous pathway, such as from hands to chest, is highly likely to be established.

It is important to note that the location must be both conducting *and* restrictive to be recognized as a conducting location with restricted movements. This chapter examines the unique hazards present in these environments and discusses the appropriate electrical safety measures that should be implemented to protect workers' well-being and prevent electrical incidents.

4.2 Automatic Disconnection of Supply

The automatic disconnection of supply requires that, in the event of failure of the basic insulation of equipment, a protective device must promptly and automatically interrupt the power supply. This disconnection should occur within a maximum timeframe, which is determined by the type of electrical system (either TT or TN) and the nominal line-to-ground voltage U_0. Figure 4.1 shows the maximum disconnection times in TN and TT systems for final circuits not exceeding 32 A and for U_0 ranging between 120 V and 230 V, based on standard IEC 60364-4-41[3].

A TN system is an electrical configuration where the neutral point of the power supply is directly grounded, and the exposed-conductive parts are connected to this ground. In contrast, a TT system is one where the exposed-conductive parts of the installation are grounded independently of the power supply ground.

---

[3]IEC 60364-4-41: "Low-Voltage Electrical Installations - Part 4-41: Protection for Safety - Protection Against Electric Shock."

Figure 4.1: Maximum disconnection time in TN and TT systems for final circuits not exceeding 32 A.

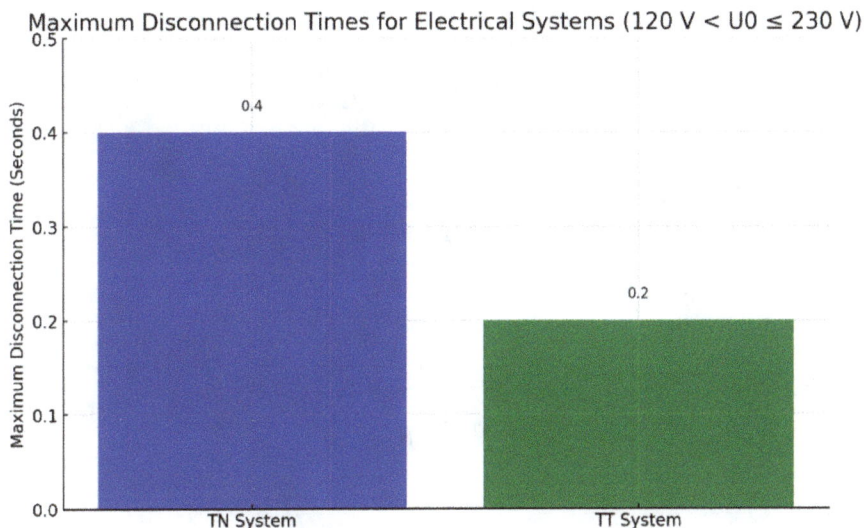

Maximum Disconnection Times for Electrical Systems (120 V < U0 ≤ 230 V)

The durations of electric shock exposure permitted by the disconnection times shown in Figure 4.1 are considered safe but only if two resistances are present in series to limit the flow of body current. The first resistance is that of the person's body R_B, and the second is the resistance between the person's body and the ground R_{BG}. In special locations where either R_B or R_{BG} may be compromised, lower disconnection times may be required. This will be discussed in the following sections.

4.2.1 Person's body resistance

The human body can be modeled as a four-terminal circuit, where the terminals represent the resistances of upper and lower limbs. The trunk resistance is generally ignored due to its large cross-sectional area and the presence of conductive fluids, resulting in a relatively low value (Figure 4.2).

In this model, the resistances R_l of the limbs are the primary factors determining the total person's body resistance R_B. In common settings such as dwelling units, the body resistance correlates with the current pathway that

Figure 4.2: Model of human body as a four-terminal circuit.

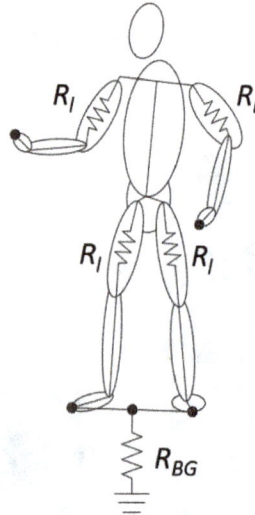

extends from both hands to both feet. In this configuration, the arms (in parallel) are in series with the legs (also in parallel), aligning with a scenario where an individual is standing and comes into contact with an electrified object. Calculations of the total body resistance can be performed using resistance values that are not exceeded for 50% of the population at 200 V, as found in relevant studies[4].

In ordinary locations, the resistance of the limbs can be calculated considering a person in dry conditions, with medium contact surface areas for the hands (i.e., 10 cm^2), and large contact surface areas for the feet (i.e., 100 cm^2). Under these conditions, the total body resistance is determined to be 741 Ω[5], as the resistance of the limbs dominates the overall resistance. This model and the corresponding resistance values provide a clear understanding of how the human body interacts with electrical currents in various scenarios.

---

[4]IEC TS 60479-1:2016, "Effects of current on human beings and livestock – Part 1: General aspects."

[5]M. Mitolo, F. Freschi and M. Tartaglia, "To Bond or Not to Bond: That is the Question," in IEEE Transactions on Industry Applications, vol. 47, no. 2, pp. 989-995, March-April 2011.

4.2.2 Person's body resistance-to-ground

The safe disconnection times also require accounting for the additional series resistance R_{BG} (Figure 4.2), which represents the person's body resistance-to-ground and is determined by the floor. For ordinary locations, R_{BG} can typically be quantified as 1 kΩ. This estimation arises from the fact that even in the absence of a floor, the resistance-to-ground of the feet, considered as parallel ground-electrodes, remains present. In ordinary locations, the resistance of footwear is not considered, and it is conservatively assumed that the person is shoeless.

In conducting locations with restricted movements, the above values of R_B and R_{BG} are likely not met, as both the body resistance and the body resistance-to-ground may have lower values. Consequently, to ensure the safety of the protective measure of disconnection of supply, shorter disconnection times than those shown in Figure 4.1 are warranted.

4.3 Conducting Location with Restricted Movements

As discussed in Chapter 2, a floor is classified as *non-conducting* if its measured resistance-to-ground is at least 50 kΩ for systems operating at voltages up to 500 V. A non-conducting floor limits the body current to safe levels, eliminating the need for any fault protection. Lower values of the floor's resistance-to-ground require the protection of persons against the risk of electric shock. If the resistance is not lower than 1 kΩ, the location can be classified as *ordinary*, rather than non-conducting, and safeguarded against indirect contact by automatic disconnection of the supply. However, if the measured resistance-to-ground of the floor is less than 1 kΩ, the location becomes *conducting*.

The chart of Figure 4.3 illustrates the resistance values for non-conducting, ordinary, and conducting locations, represented on a logarithmic scale for better visualization.

In this context, the type of conductive material present at the location is irrelevant; the only factor that matters in terms of electric shock hazard is its low resistance-to-ground.

A person's movements become constrained within a space when its dimensions are comparable to their body size. In such situations, the current pathway from hands to chest is very likely, as illustrated in Figure 4.4.

Figure 4.3: Values of resistance-to-ground for different types of locations.

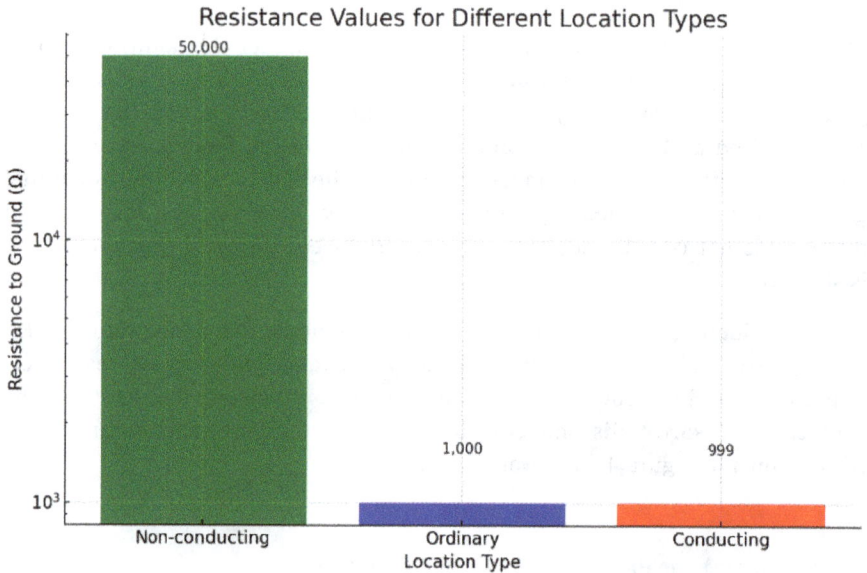

Resistance Values for Different Location Types

Figure 4.4: Current pathway from hands to chest.

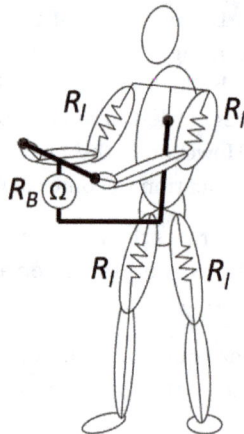

In these conditions, the upper limbs are in parallel, the lower limbs are excluded, and the total body resistance becomes $R_B = \frac{R_l}{2}$. It is important to note that conducting restrictive locations may be water wet, which can significantly reduce the limb resistances and, consequently, cause a hazardous increase in body current for a given touch voltage.

The probability of ventricular fibrillation is impacted by the amount of current passing through the heart, which varies based on the path the current takes. The heart–current factor F, defined in technical standard IEC TS 60479-1, quantifies this probability based on investigations of electrical injuries and from measurements on cadavers. A higher F value indicates a more hazardous current pathway. The standard indicates that the hands-to-chest pathway, for which $F = 1.5$, is more hazardous than the both-hands-to-both-feet pathway, for which $F = 1$. For instance, a 66.7 mA hands-to-chest current poses the same ventricular fibrillation risk as a 100 mA left-hand-to-both-feet body current. Furthermore, simulations[6] have established an even more hazardous scenario, where a current of just 17.5 mA through the chest can cause ventricular fibrillation with the same probability as a 100 mA current for the both-hands-to-both-feet path.

As shown in Figure 4.4, the hands-to-chest pathway does not include the person's body resistance-to-ground R_{BG}, which eliminates the beneficial limiting effect of an additional series resistance on the body current.

4.4 Fixed Electric Equipment

Fixed electric equipment refers to equipment fastened to a support or otherwise secured in a specific location, such as lighting fixtures permanently attached to ceilings or walls. This type of equipment is permitted in conducting locations with restricted movements, provided certain conditions are met, as discussed in the following sections.

4.4.1 Class I fixed equipment

As previously discussed, Class I equipment features the basic insulation of live parts and the protective equipotential bonding connection, which protects against faults in conjunction with a protective device by automatically disconnecting the power supply. Class I fixed equipment is allowed in CLRs

---

[6]M. Mitolo, M. Tartaglia, F. Freschi, A. Guerrisi: "Numerical Simulation of Heart-Current Factors and Electrical Models of the Human Body". IEEE Transactions on Industry Applications; Volume 49, Issue 5, September/October 2013, pp. 2290-2299.

Figure 4.5: Automatic disconnection of supply in conjunction with supplementary protective equipotential bonding (EQS).

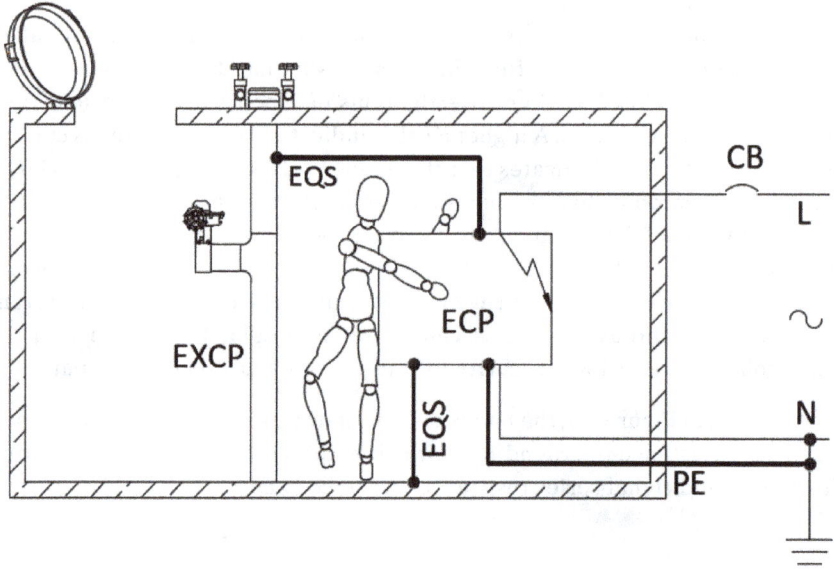

provided that a supplementary protective equipotential bonding (EQS) is in place (Figure 4.5).

The supplementary bonding must connect all the exposed-conductive parts (ECPs) of the fixed equipment to the extraneous-conductive parts (EXCPs), for example, the metalwork of a boiler or the metal floor. This supplementary equipotential bonding is crucial in reducing the potential difference between a faulty piece of equipment and all the EXCPs within the conducting location.

4.4.2 Class II fixed equipment

Class II fixed equipment is permitted in CLRs. However, according to IEC 60364-7-706, this measure must be employed in conjunction with additional protection provided by residual current devices (RCDs), known as ground fault circuit interrupters (GFCIs) in the US, which monitor the circuits supplying power to these locations. The residual current represents the algebraic sum of the electric currents in all live wires, including the neutral, at a given point in an electric circuit, which is zero under normal conditions. RCDs are designed to promptly interrupt the supply when the residual current is non-zero and exceeds their ratings. For CLRs, the RCDs should have a rated residual current not exceeding 30 mA.

4.4.3 Electrical separation

Electrical separation can also serve as a protective measure for fixed electric equipment and is also applicable for hand-held and mobile electrical equipment. Given the higher risk in CLRs, electrical separation must be achieved using an isolating transformer with double (or reinforced) insulation between windings, rather than just basic insulation. These transformers, characterized by identical primary and secondary voltages (e.g., 120 V/120 V), galvanically separate the equipment from the power source, preventing current flow in the event of contact with live parts.

Under this measure, only one piece of equipment should be connected to the isolating transformer's secondary winding. If multiple devices are powered by a single transformer, a hazardous situation can occur if two ground-faults involve both secondary poles of the transformer (Figure 4.6).

In such a scenario, workers can be exposed to a hazardous potential difference if they simultaneously contact two faulty ECPs. These ground-faults can go undetected, creating an environment where hazardous potential differences

Figure 4.6: Two pieces of equipment supplied by a single separation transformer in the event of ground-faults.

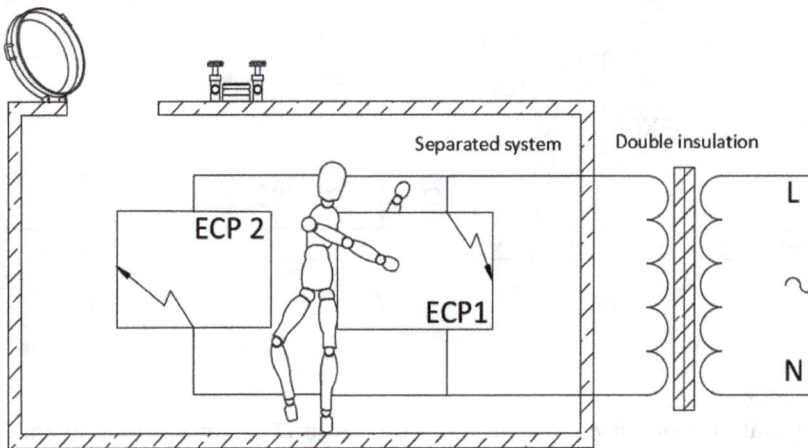

exist between the equipment casings. This could persist indefinitely because the electrical separation measure does not require any insulation monitoring device that could trigger an audible or visual alarm at the first ground-fault[7]. Therefore, receptacle outlets in CLRs are critical, as unaware users may use power strips to supply multiple devices from the same transformer.

4.4.4 Safety extra-low voltage (SELV) and protective extra-low voltage (PELV)

Another protection strategy for fixed equipment consists of the safety extra-low voltage (SELV) systems, where, as previously discussed, the operating voltage, provided with safety isolating transformers, is not allowed to exceed a maximum of 50 V a.c. and 120 V d.c. SELV circuits may also be used for hand-held and mobile electrical equipment. Protective extra-low voltage (PELV) circuits are also permitted in CLRs, provided that equipotential bonding is in place. This bonding should link all ECPs and EXCPs of the location, along with the grounding terminal of the PELV supply (Figure 4.7).

Figure 4.7: PELV system with equipotential bonding.

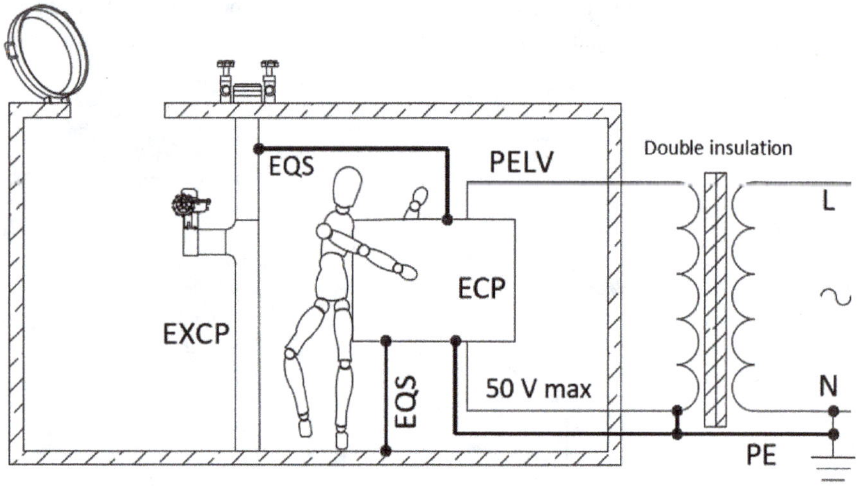

[7]Insulation monitoring devices are required in IT systems, where live parts are insulated from ground or connected to ground through a sufficiently high impedance.

Sources for SELV and PELV should be located outside the CLR unless they are part of the fixed equipment.

It is important to note that PELV systems for conductive restrictive locations are not universally accepted. For instance, they are not permitted in Italy and France.

4.5 Hand-held Equipment

Electric equipment intended to be held in the hand during normal use is typically supplied via a flexible cord connected to a receptacle. While in use in the CLR, the cord is trailed, potentially leading to abrasion, twisting, and straining, which makes it susceptible to damage. This increases the risk of electric shock hazards for workers. Technical standards stipulate that for supplies to hand-held tools the only permissible protective measures are electrical separation with an isolating transformer or SELV circuits. In both cases, the source must be located outside the CLR to effectively mitigate the risks. Hand-lamps in CLRs are considered particularly hazardous in terms of direct contact, as workers may attempt to replace the lamp without first disconnecting the supply. To address this risk, the only permissible safety measure for handlamps is the use of SELV circuits. This approach ensures that the handlamps operate at extra-low voltage, thereby reducing the risk of electric shock during maintenance.

Although not a mandatory requirement, the use of Class II hand-held equipment in CLRs it is always recommended. Figure 4.8 shows a Class II drill that is powered via an isolating transformer (e.g., 120 V/120 V) providing the same safety level as handheld equipment powered by SELV circuits.

An electric shock could occur only if the double insulation of both the isolating transformer and the equipment were to fail. This situation would involve a total of four faults occurring, as shown in Figure 4.8. Under these circumstances, the supply source to the hand tool would become grounded, and if a worker were to hold it, an electric shock could occur. As previously discussed, the probability of occurrence of these faults is deemed extremely low.

With the SELV supply, being at a voltage of 50 V or less, even in the unlikely presence of the four faults, no risk of electric shock would be present. This highlights the critical safety advantages of using SELV systems in CLRs. SELV systems ensure that even under multiple fault conditions, the risk of electric shock is mitigated due to the inherently safe operating voltage. The reliability of SELV circuits in maintaining safety, even in the event of faults, makes them

Figure 4.8: Separated system and Class II hand-held equipment.

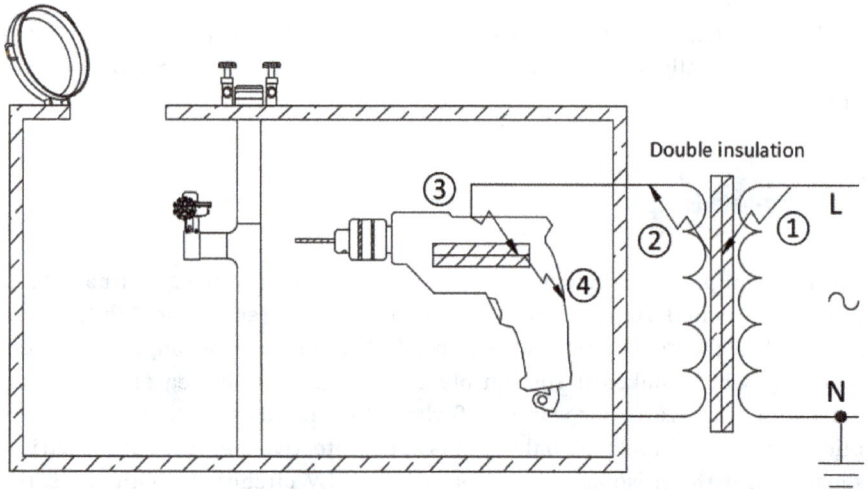

Double insulation

the preferred choice for electrical safety in environments with heightened risk factors, such as CLRs.

The use of a 30 mA RCD is not required where the source of supply originates from a circuit that is protected by SELV electrical separation.

It is worth noting that residual current devices employed with either an isolating transformer or a safety transformer do not provide any additional safety because they can never operate as intended in the event of ground-faults, as shown in Figure 4.9.

Figure 4.9: (a) RCD installed on the primary side; (b) RCD installed on the secondary side.

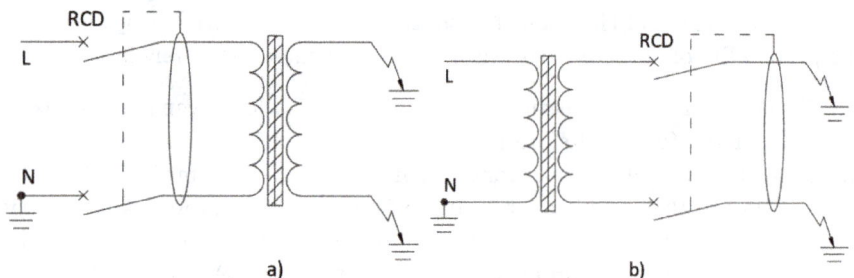

The primary function of an RCD is to detect imbalances between the live and neutral conductors, which typically indicate a current leakage to the ground. However, in the case of an isolating transformer, the secondary side is electrically separated from the primary side. Therefore, a double ground-fault on the secondary side does not create a residual current that an RCD on the primary side can detect (Figure 4.9a). Similarly, installing an RCD on the secondary side of the transformer does not improve safety. A double ground-fault does not create a differential current between live and neutral conductors that the RCD can detect, and thus can never trip (Figure 4.9b).

4.6 Summary

This chapter provides a comprehensive overview of conducting locations with restricted movements (CLRs) as defined by IEC 60364-7-706. CLRs are areas primarily enclosed by extraneous-conductive parts (EXCPs), which are conductive elements not part of the electrical installation but likely to introduce dangerous potentials into the location. Typical examples include transmission towers, metal tanks, and damp tunnels. These environments pose significant hazards due to the likelihood of extensive bodily contact with grounded conductive materials, which reduces the body's resistance to ground and increases the risk of electric shock, potentially leading to life-threatening conditions like ventricular fibrillation. The chapter details the specific dangers of electrical equipment in CLRs, noting the various potential current pathways and their associated risks. It also outlines the deficiencies of the standard automatic disconnection of supply in case of faults, describing the operational differences and safety requirements for CLRs. The human body's resistance modeled as a four-terminal circuit is discussed, emphasizing the lower resistance values in CLRs due to possible wet conditions. Safety measures for fixed and hand-held equipment in CLRs are thoroughly examined, including the use of Class I and II equipment, electrical separation with isolating transformers, and safety extra-low voltage (SELV) systems. The chapter highlights the limitations of residual current devices (RCDs) in CLRs, explaining why they cannot provide additional safety when used with isolating or safety transformers. The importance of supplementary equipotential bonding in Class I equipment and the recommendations for Class II and SELV systems are emphasized to ensure the protection of workers in these high-risk environments.

Key Definitions

Automatic disconnection of supply: A safety measure that requires the power supply to be automatically interrupted in the event of a fault between the line conductor and an exposed-conductive part of the equipment. The disconnection occurs within a safe maximum timeframe to prevent electric shock.

Class I fixed equipment: Electrical equipment with basic insulation of live parts and a protective equipotential bonding connection. This class of equipment relies on automatic disconnection of the power supply and supplementary protective equipotential bonding in conductive locations with restricted movements (CLRs).

Class II fixed equipment: Electrical equipment that does not rely on a protective ground connection and instead uses double or reinforced insulation to protect against electric shock. In CLRs, it must be used with additional protection provided by residual current devices (RCDs).

Conducting locations with restricted movements (CLRs): Areas predominantly enclosed by extraneous-conductive parts (EXCPs), where movement is restricted due to the location's dimensions or working conditions. These environments pose significant electric shock risks due to extended bodily contact with grounded conductive materials.

Electrical separation: A protective measure where an isolating transformer with double insulation separates the equipment in a CLR from the power source, preventing current flow in the event of contact with live parts. This measure is applicable to hand-held and mobile electrical equipment in CLRs.

Extraneous-conductive parts (EXCPs): Conductive elements that are not part of the electrical installation but can introduce dangerous potentials or the zero potential into a location. These parts are commonly found in CLRs and can pose significant electric shock risks.

Fixed electric equipment: Electrical equipment secured in a specific location, such as lighting fixtures attached to ceilings or walls. This type of equipment is permitted in CLRs under certain conditions, including appropriate protective measures.

Hand-held equipment: Electric tools and devices intended to be held in the hand during normal use. In CLRs, such equipment must be exclusively supplied via electrical separation with an isolating transformer or safety extra-low voltage (SELV) circuits, with the power source located outside the CLR.

Heart-current factor F: A factor defined in IEC TS 60479-1 that quantifies the probability of ventricular fibrillation based on the path the electric current takes through the body. A higher F value indicates a more hazardous current pathway, such as the hands-to-chest route.

Person's body resistance: The resistance of the human body to electric current, modeled as a four-terminal circuit. The resistance of the upper and lower limbs (R_l) determines the total body resistance (R_B), which varies based on contact surface area and environmental conditions.

Person's body resistance-to-ground: The resistance between a person's body and the ground, influenced by the floor covering. In ordinary locations, it is typically quantified as 1 kΩ, but in CLRs, this resistance can be significantly lower.

Residual current devices (RCDs): Safety devices that detect imbalances between live and neutral conductors, indicating current leakage to the ground. RCDs interrupt the power supply when the residual current exceeds their ratings. In CLRs, they provide additional protection for Class II fixed equipment.

Safety extra-low voltage (SELV): A protective measure involving circuits with a supply voltage not exceeding 50 V a.c. or 120 V d.c. SELV systems are used for both fixed and hand-held equipment in CLRs, ensuring safety even under fault conditions due to the inherently safe operating voltage.

Trunk resistance: The resistance of the human torso, generally considered low due to its large cross-sectional area and conductive fluids. In body resistance models, trunk resistance is often ignored, focusing instead on the resistance of the limbs.

Bibliography

1. IEC 60364-7-706:2024, "Low-voltage electrical installations - Part 7-706: Requirements for special installations or locations - Conducting locations with restricted movement."
2. M. Mitolo: "Protecting Electrical Workers in Conducting Locations with Restricted Movements," Distributed Generation & Alternative Energy Journal (DGAEJ). Vol. 39_4, pp. 1–12.
3. "Restrictive Conductive Location Definition." Law Insider, www.lawinsider.com/dictionary/restrictive-conductive-location.AccessedJune2024.
4. IEC 60364-4-41:2005: "Low-Voltage Electrical Installations - Part 4-41: Protection for Safety - Protection Against Electric Shock."
5. M. Mitolo, M. Tartaglia and S. Panetta, "Of International Terminology and Wiring Methods Used in the Matter of Bonding and Earthing of Low-Voltage Power Systems," in IEEE Transactions on Industry Applications, vol. 46, no. 3, pp. 1089-1095, May-June 2010.
6. IEC TS 60479-1:2016, "Effects of current on human beings and livestock – Part 1: General aspects."

7. M. Mitolo, F. Freschi and M. Tartaglia, "To Bond or Not to Bond: That is the Question," in IEEE Transactions on Industry Applications, vol. 47, no. 2, pp. 989-995, March-April 2011.

8. IEC 60364-6:2016, "Low voltage electrical installations - Part 6: Verification."

9. M. Mitolo, M. Tartaglia, F. Freschi, A. Guerrisi: "Numerical Simulation of Heart-Current Factors and Electrical Models of the Human Body". IEEE Transactions on Industry Applications; Volume 49, Issue 5, September/October 2013, pp. 2290-2299.

10. IEC 60050-195:2021, "International Electrotechnical Vocabulary (IEV) - Part 195: Earthing and protection against electric shock."

11. IEC 60050-826:2022, "International Electrotechnical Vocabulary (IEV) - Part 826: Electrical installations."

12. IEC 61558-2-4:2021, "Safety of transformers, reactors, power supply units and combinations thereof - Part 2-4: Particular requirements and tests for isolating transformers and power supply units incorporating isolating transformers for general applications."

13. M. Mitolo: "Protecting Electrical Workers in Conducting Locations with Restricted Movements," Distributed Generation & Alternative Energy Journal (DGAEJ). Vol. 39, Issue 4, 1-12. DOI: 10.13052/dgaej2156-3306.3944. River Publishers

5

Electrical Safety Engineering of PV Systems

5.1 Introduction

Photovoltaic (PV) systems, whether ground-mounted or rooftop installations, typically utilize metal racks, frames, and mounting structures to support the modules. This brings forth the critical issue of whether equipotential bonding of these conductive parts is necessary for safety. Equipotential bonding aims to ensure that all conductive parts of an installation maintain the same electrical potential, thereby mitigating the risk of electric shocks in the event of a fault.

Some PV systems may be equipped with solar trackers that use servo motors to orient the panels towards the sun, optimizing energy production. If the basic insulation of a servo motor fails, the mounting structures could become energized, underscoring the importance of proper bonding practices.

As discussed in previous chapters, only exposed-conductive parts (ECP) and extraneous-conductive parts (EXCP) should be bonded to ensure safety. This chapter presents criteria for determining whether conductive parts in a PV system fall in these categories, based on the specific characteristics of the installation.

5.2 Safety Aspects of Photovoltaic Module Frames

An integral part of photovoltaic (PV) modules is their frame, which provides essential structural support and protection, and facilitates the installation process. These frames are commonly constructed from extruded aluminum profiles. PV modules are designed with basic insulation to prevent electric shock by isolating live electrical components from accessible parts, such as the frame. However, because the frame can be touched and may become live if the basic insulation fails, the PV module is classified as an exposed-conductive part (ECP). To ensure safety and enable protective devices to function correctly in the event of a fault, the PV module should be connected to a grounding system via a protective conductor (PE) (Figure 5.1).

Figure 5.1: Exposed-conductive part and extraneous-conductive part in PV systems.

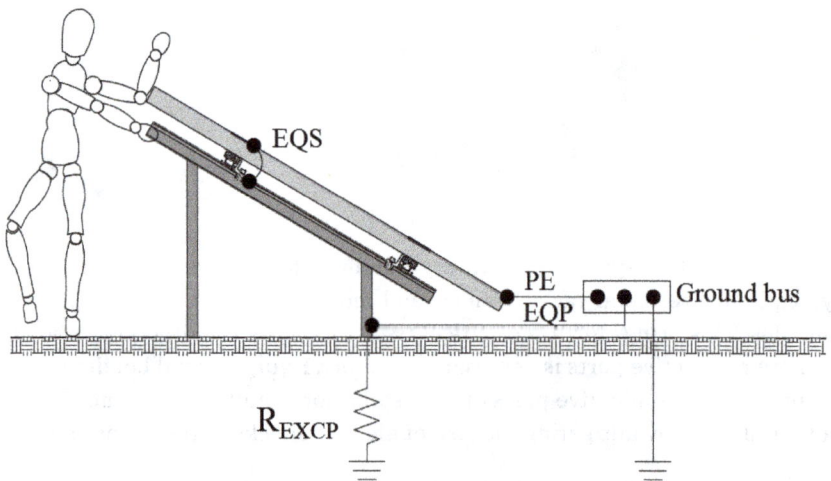

5.3 Classifying and Bonding Mounting Structures as EXCPs in PV Systems

As previously discussed, extraneous-conductive parts (EXCPs) are conductive elements that can be touched and can introduce either zero potential or an arbitrary potential. Mounting structures of PV systems may fall into this category if they bring the zero potential through their ground-resistance R_{EXCP} (Figure 5.1). Under a ground-fault condition of the PV module, a person may simultaneously contact the faulted module and the mounting structure, being exposed to the highest touch voltage.

Mitigating this hazardous situation involves the equipotential bonding of the mounting structures to the ground terminal. However, it is crucial to identify the mounting structures as EXCPs before proceeding with their electrical bonding. Measuring the ground-resistance R_{EXCP} is part of the process of this classification (Figure 5.2). The test to measure R_{EXCP} should be conducted utilizing a continuity tester equipped with a voltage range between 4 and 24 V and employing a test current of 200 mA.

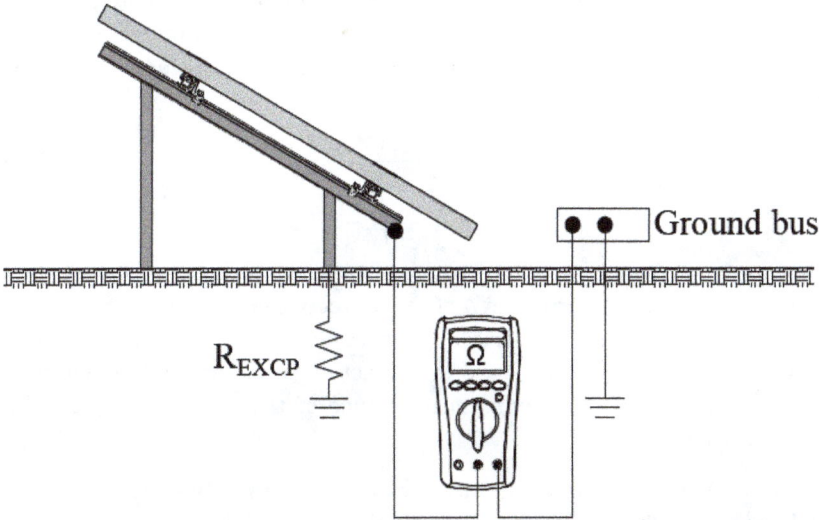

Figure 5.2: Resistance test to identify mounting structures as EXCPs.

If the measured resistance exceeds the value calculated in Equation (5.1), then the mounting structure does not need to be connected to the ground terminal.

$$R_{EXCP} = \frac{U_0}{I_B} - R_B. \tag{5.1}$$

U_0 represents the nominal voltage to ground of the PV installation; R_B is the total body impedance for a current path hand-to-hand, applicable for large surface areas of contact in water-wet conditions, which is not exceeded for 95%

of the population[1]; I_B denotes the value of the body current that should not surpass dangerous limits. For instance, considering $U_0 = 500$ V, $I_B = 30$ mA (the threshold current causing reversible disturbances of heart function, but typically resulting in no organic damage), and $R_B = 1150\,\Omega$ (value corresponding to $U_0 = 500$ V for both a.c. 50/60 Hz and d.c. voltages), the value of R_{EXCP} is calculated to be 15.5 kΩ. Mounting structures with a resistance-to-ground less than this value should be bonded. It is important to note the difference between this value and the less stringent result obtained in dry conditions calculated for R_{EXCP} in Chapter 1, which was 10.9 kΩ.

If the mounting structure is identified as an EXCP, in addition to equipotential bonding, supplementary protective equipotential bonding (denoted as EQS in Figure 5.1) is necessary to further mitigate the risk of electric shock. This connection is implemented between the mounting structures and the PV module frames to ensure they maintain the same potential. When there is uncertainty regarding the effectiveness of the supplementary bonding jumper, it must be verified that its resistance R meets the following conditions:

$$R \leq \frac{120\,V_{dc}}{I_a} \tag{5.2}$$

$$R \leq \frac{50\,V_{ac}}{I_a}. \tag{5.3}$$

I_a is the operating current in A of the protective device, which can be either the rated residual current for RCDs or the 5 s operating current for overcurrent devices. For example, in the case of protection in a.c. via a residual current device rated 30 mA the maximum resistance of the bonding jumper cannot exceed 1.6 kΩ.

Unlike land-based installations, the mounting racks on rooftops are generally not considered extraneous-conductive parts. They are, in fact, in good electrical contact with the building itself and are therefore grounded through its resistance R_{BLDG} (Figure 5.3)

This result remains consistent regardless of the actual resistance of the building, as no current circulates through it under ground-fault conditions. Consequently, in the event of insulation failure of the PV module, the mounting structures remain at the same electrical potential as that of the PV equipment, making additional equipotential bonding unnecessary.

---

[1] IEC 60479-1 "Effects of current on human beings and livestock."

Figure 5.3: Rooftop-mounted racks and metal frames of PV modules are equipotential in the event of ground-faults.

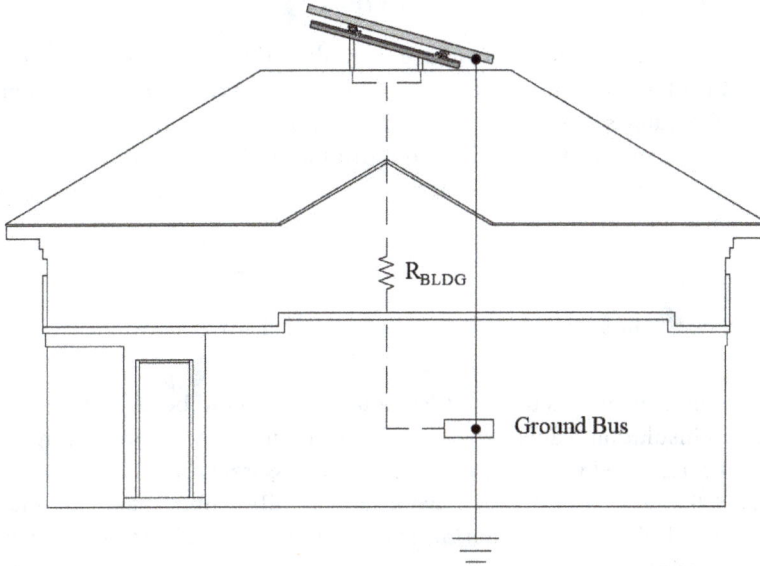

5.4 Grounding and Bonding Considerations for Class I PV Modules and Mounting Structures

PV modules and inverters can be classified as Class I, indicating that they have basic insulation and connection of the protective conductor to the grounding terminal of the installation. In the event of basic insulation failure, the protective conductor allows fault current to flow, enabling the PV ground-fault protection system to disconnect and isolate the faulted part of the PV array from the rest of the system.

If there is good electrical continuity between the PV module and the supporting structure (e.g., resistance less than 0.1Ω), the mounting rack can serve as the bonding connection between the modules. This approach simplifies the installation process and reduces the need for numerous bonding jumpers. In this case, the mounting rack system must be listed, labeled, and identified for this purpose and thoroughly evaluated to ensure that its metal structure maintains electrical continuity and mechanical integrity throughout its span.

It is not necessary to ground the mounting structure merely because it is in contact with the PV module. As previously mentioned, a conductive part that can be energized simply because it is in contact with Class I equipment is not classified as an exposed-conductive part (ECP).

However, the mounting structure may directly support single-core non-sheathed cables (i.e., Class I cables), or be in contact with wireways housing them. If the cable's basic insulation fails, the supporting rack becomes energized. In this scenario, the structure qualifies as an ECP and must be included in the equipotential bonding system, even though it is not classified as an extraneous-conductive part.

5.5 Bonding Requirements for Class II PV Equipment

Current requirements dictate that PV equipment should be Class II or possess equivalent insulation. Cables, whether single or multi-conductor, are considered Class II if their rated voltage is not less than the system's nominal voltage (and at least 300/500 V) and if they have a non-metallic outer sheath in addition to basic insulation. The insulating sheath prevents equipment energization should the primary insulation fail. Therefore, Class II PV modules should not be connected to a protective conductor. Mounting structures of Class II PV modules do not need to be bonded, even if classified as extraneous-conductive parts. This is because the supplementary insulation effectively separates a person in contact with the rack from energized parts in the event of failure of the basic insulation.

However, metal frames of Class II modules may still include terminals for functional grounding, which is unrelated to safety. They can be connected to ground through an electronic means, typically internal to the inverter or charge controller, which provides the module insulation monitoring. The insulation resistance R_{iso} between live parts and the ground can degrade due to wet operating conditions or aging of the insulation material. When it falls below a certain threshold, the monitoring devices can shut down the inverters. Figure 5.4 provides the minimum insulation resistance thresholds for detecting insulation failure to ground relative to the PV array power rating.

Figure 5.4: Minimum R_{iso} values versus PV array power rating.

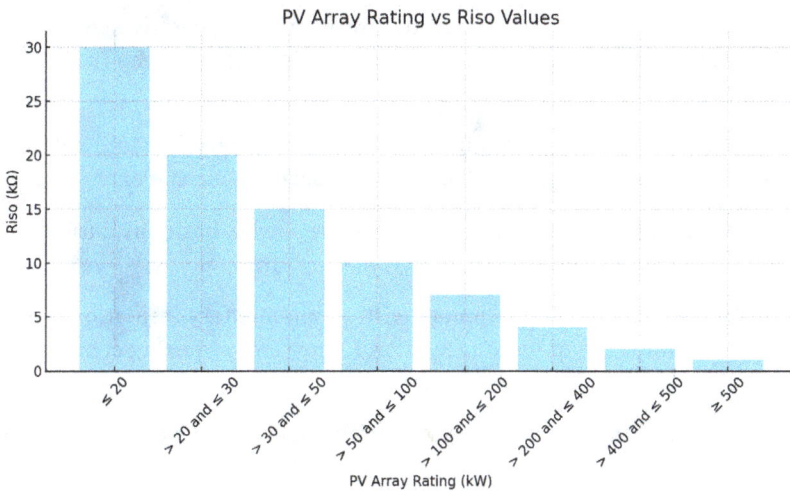

5.6 Summary

This chapter addresses the need for equipotential bonding in photovoltaic (PV) systems, which can be installed either on the ground or rooftops. PV systems use metal racks, frames, and mounting structures, raising safety concerns about whether these conductive parts require bonding. Equipotential bonding ensures all conductive parts are at the same electrical potential, reducing electric shock risks in the event of faults.

Exposed-conductive parts (ECPs) include PV module frames, which, due to basic insulation failure, can become live and must be connected to a grounding system. On the other hand, extraneous-conductive parts (EXCPs) are conductive elements like mounting structures that can introduce a zero or arbitrary potential. They must be assessed for ground-resistance R_{EXCP} to determine if bonding is necessary.

For Class I equipment, which includes PV modules and inverters with basic insulation, the mounting rack can serve as the bonding connection if they provide good electrical continuity. The chapter also discusses how mounting racks on rooftops generally do not need to be bonded as they are grounded through the building's resistance. Finally, the monitoring of insulation resistance R_{iso} is crucial for detecting insulation degradation and ensuring the safe operation of PV systems; this requires a ground connection also of Class II PV modules.

Key Definitions

Basic insulation: Insulation designed to protect against electric shock by separating live electrical components from accessible parts, like the frame of PV modules.

Basic insulation failure: A condition where the primary insulating layer of electrical equipment fails, causing conductive parts to become energized.

Class I PV equipment: PV modules and inverters with basic insulation and a protective conductor connected to the grounding terminal.

Class II PV equipment: PV equipment with supplementary insulation preventing energization in case of basic insulation failure, not requiring connection to a protective conductor for safety purposes.

Electrical continuity: A measure of the effectiveness of bonding connections, for example ensuring resistance less than 0.1 Ω between PV modules and supporting structures.

EXCPs in PV systems: Conductive elements that can be touched and introduce zero or arbitrary potential, like mounting structures with specific ground resistance.

Insulation resistance R_{iso}: Resistance between live parts and ground, monitored to detect insulation degradation and prevent faults.

Equipotential bonding: A safety measure to ensure all conductive parts of an installation are at the same electrical potential, reducing the risk of electric shocks during faults.

Non-metallic outer sheath: An additional layer of insulation of Class II cables, providing enhanced protection against electric shock.

Photovoltaic (PV) systems: Systems that convert sunlight into electricity, installed either on the ground or rooftops, typically using metal racks, frames, and mounting structures.

PV module frame: Structural support for PV modules, typically made of extruded aluminum profiles, providing protection and facilitating installation.

R_{EXCP}: Resistance through which an EXCP is naturally grounded, measured to determine the need for electrical bonding.

Single-core non-sheathed cable: Class I cable.

Solar trackers: Devices equipped with servo motors to orient PV panels toward the sun, enhancing energy production.

Bibliography

1. M. Mitolo, F. Freschi and M. Tartaglia, "To Bond or Not to Bond: That is the Question," *IEEE Transactions on Industry Applications*, vol. 47, no. 2, pp. 989-995, March-April 2011.
2. IEEE Std 3003.2 – 2014: "Recommended Practice for Equipment Grounding and Bonding in Industrial and Commercial Power Systems."
3. Low-Voltage Electrical Installations - Part 4-41: Protection for Safety - Protection Against Electric Shock, IEC 60364-4-41, 2005.
4. The National Electrical Code, NFPA 70, 2023.
5. IEC 60479-1:2018: "Effects of current on human beings and livestock."
6. IET Guidance Note 8: "Earthing & Bonding," 5th Ed., 2022.
7. IEC 60364-7-712: "Low voltage electrical installations – Part 7-712: Requirements for special installations or locations – Solar photovoltaic (PV) power supply systems." 2017.
8. M. Mitolo, F. Freschi, R. Tommasini: "Electrical Model of Building Structures under Ground-Fault Conditions, Part I", *IEEE Transactions on Industry Applications*, Vol. 52, No. 2; March/April 2016.
9. BS 7671:2018 "Requirements for Electrical Installations," *IET Wiring Regulations*, 18th Ed.
10. R. Araneo and M. Mitolo, "Insulation Resistance and Failures of a High-Power Grid-Connected Photovoltaic Installation: A Case Study," *IEEE Industry Applications Magazine*, vol. 27, no. 3, pp. 16-22, May-June 2021.
11. IEC 62548-1 Photovoltaic (PV) arrays - Part 1: Design requirements
12. IEC 61730-1:2023 Photovoltaic (PV) module safety qualification - Part 1: Requirements for construction.
13. IEC 61140: 2016, "Protection against electric shock – Common aspects for installation and equipment."
14. M. Mitolo, R. Araneo: "Equipotential Bonding of Photovoltaic Systems," I&CPS Europe 2024.

Electrical Safety Engineering for the Mitigation of Electrostatic Hazards

6.1 Introduction

Static charge is generated through various mechanisms, predominantly involving the separation of positive and negative charges between two unlike materials. When two distinct materials come into contact and subsequently separate, electrons can migrate from one surface to the other, resulting in an imbalance of charges. This phenomenon, known as *triboelectric charging*, is influenced by several factors including the intrinsic properties of the materials involved, the extent of their contact surface area, and the environmental conditions in which the interaction occurs.

Material properties play a crucial role in static charge generation. Different materials exhibit varying tendencies to gain or lose electrons, a characteristic defined by their position in the triboelectric series. For instance, materials such as glass and nylon tend to acquire positive charges, while materials like Teflon and silicon tend to acquire negative charges. This disparity is attributed to the electron affinity of the materials, which determines how readily they accept or donate electrons during contact and separation.

Surface area is another significant factor. The larger the contact area between two materials, the greater the potential for charge transfer. This is because a larger surface area provides more opportunity for electron exchange, thus amplifying the resulting static charge. Additionally, the nature of the surface, whether it is smooth or rough, can impact the effectiveness of charge transfer, with smoother surfaces typically facilitating more uniform electron movement.

Environmental conditions, including humidity and temperature, also significantly influence static charge production. Higher humidity levels tend to reduce static charge accumulation as moisture in the air can facilitate the dissipation of charges. Conversely, in dry conditions, static charges are more likely to build up and persist. Temperature variations can affect the conductivity of materials and the rate of electron movement, further impacting the generation and stability of static charges.

Static charge, once generated, can accumulate on both conductive and insulating objects, as well as in high-resistivity liquids and gases. This accumulation leads to potential differences between objects and between objects and the ground, which can reach up to 100 kV. Such high potential differences pose significant risks in various industrial and work environments.

Grounding is the most effective method for mitigating these hazards. By grounding conductive parts that could reach hazardous charge levels, charge accumulation is prevented, and the risk of energy release as a single spark, either to the ground or another object, is minimized. This practice ensures that any static charge is safely dissipated, reducing the likelihood of ignition of flammable substances and the risk of electric shock.

Incorporating these measures into industrial protocols can substantially reduce the risks associated with static electricity, leading to safer working environments.

6.2 Classification and Behavior of Solid Materials Based on Electrical Resistance

Solid materials may be classified as conductive, dissipative, or insulating, based on their electrical resistance, as shown in the chart of Figure 6.1, which is based on IEC TS 60079-32-1[1].

The chart employs a logarithmic scale to display the wide range of resistance values for these materials. It covers the resistance-to-ground for footwear and floors, surface resistance for clothes, and resistance per unit length for pipes.

---

[1]IEC TS 60079-32-1, "Explosive atmospheres – Part 32-1: Electrostatic hazards, guidance."

Figure 6.1: Classification of solid materials based on their resistance.

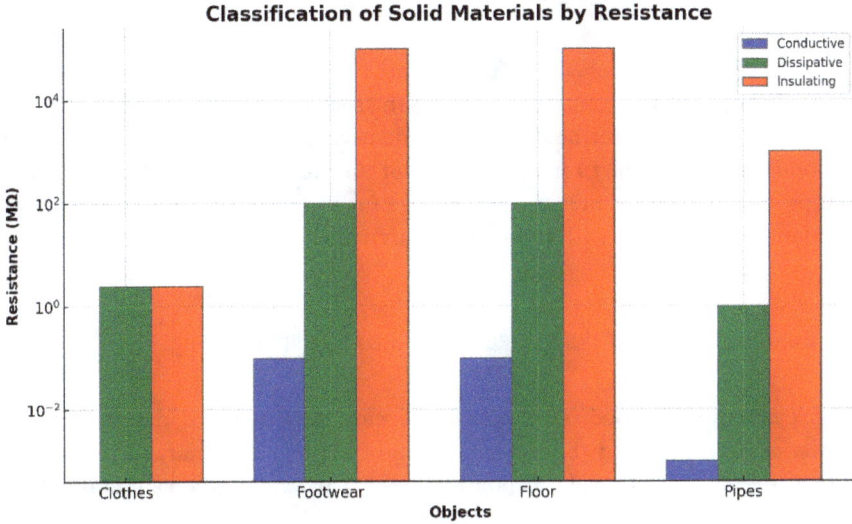

Dissipative materials function similarly to conductors in dissipating static electric charges. For example, a floor is classified as dissipative if its resistance ranges between 100 kΩ and 100 MΩ.

When two bodies come into contact, electrons are transferred from one to the other. Upon separation, if both materials are conductors, the electric charges disperse and neutralize. In contrast, an insulating material retains its electric charge due to its high resistivity. Repeated contact, such as in industrial painting processes or handling powdered substances in pharmaceutical manufacturing, can generate significant electric charge, especially across large contact areas. Transferring fuels, oils, or other flammable liquids through plastic or rubber hoses can also promote the generation of static electricity. If the parts are isolated from the ground and the charge formation continues, the electric charge accumulates over time. Charge accumulation occurs only if the rate of charge generation exceeds the rate of dissipation. If the mechanism of charge formation stops, the charge decays exponentially over time, leaking through the object's resistance-to-ground.

The duration required for static charge on a solid surface, or within the bulk of a liquid or powder, to decrease to 37% of its initial value is referred to as the *relaxation time* τ. For instance, gasoline, which has a low conductivity ranging from 10^{-13} S/m to 10^{-10} S/m, exhibits a relaxation time that varies from 0.2 s to 200 s. This wide range in relaxation time affects how quickly charge can dissipate.

Furthermore, the relaxation time tends to increase with the resistivity of the medium, as higher resistivity impedes the recombination of electric charges. For example, pure water possesses a conductivity of 5×10^{-6} S/m, which corresponds to a very short relaxation time of approximately $1\mu s$. This indicates that pure water cannot hold a static charge for a significantly extended period compared to materials with lower conductivity.

6.3 Generation and Transfer of Electric Charge in Liquid Handling

As previously discussed, electric charge can be generated due to friction between the liquid and the inner surface of a pipe. The charged liquid can, in turn, energize the vessel it is poured into. For example, when a metallic tank is being filled with a negatively charged liquid, charges of opposite signs are induced on its inner and outer surfaces (Figure 6.2).

As the charged liquid continues to flow, the induced charge increases, resulting in higher electric potentials. If the liquid has low conductivity, the charge does not dissipate quickly, leading to accumulation of static electricity. This accumulated charge can pose risks such as sparks or electrostatic discharge, which are hazardous in environments with flammable or explosive materials. To mitigate the risk, grounding the tank allows the negative charges on its outer surface to drain away. However, the charge on the inner surface and in the liquid will persist for the relaxation time.

Figure 6.2 also illustrates the equivalent electrical circuit for the electrostatically charged tank. In this circuit, R represents its ground-resistance, and C denotes its capacitance-to-ground. The instantaneous ground potential $v(t)$ is given by Equation (6.1), where I is the charging current, which describes the rate at which charge is accumulated in the tank.

$$v(t) = IR\left(1 - e^{-\frac{t}{RC}}\right). \tag{6.1}$$

The charging current I, which is influenced by the properties of the liquid flowing through pipes, depends on several factors. Specifically, for a non-conductive liquid experiencing turbulent flow, the streaming current I (in amperes) can be expressed by Equation (6.2):

$$I = Nv^x d^y, \tag{6.2}$$

Figure 6.2: Charging of a metal tank by the flow of liquid.

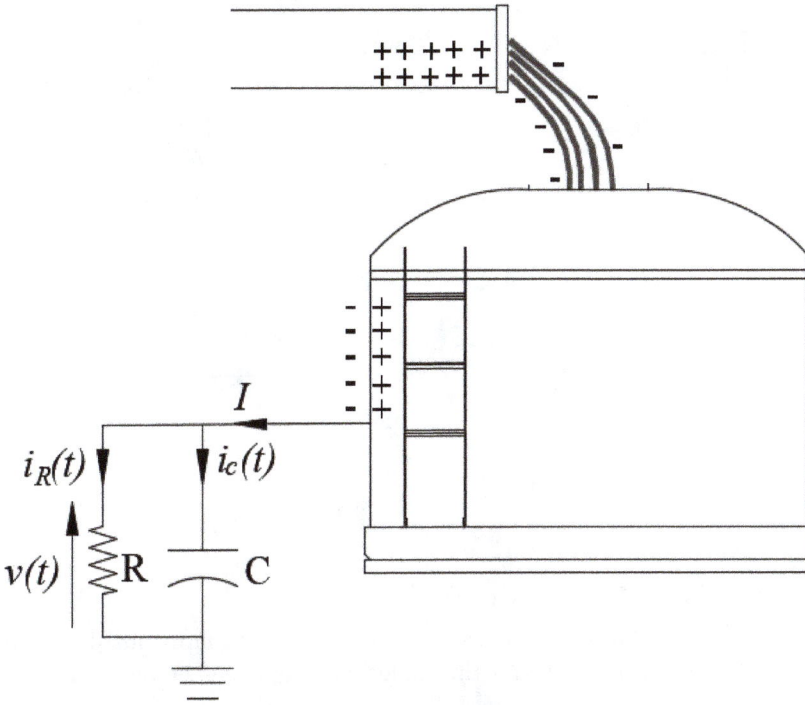

where v is the flow velocity (m s^{-1}) and d is the diameter of the pipe (m). An order-of-magnitude estimate for N (constant characterizing specific flow conditions) is 10^{-5} C s m^{-4}. It has been most commonly reported in the literature that both exponents x and y are approximately equal to 2.

The chart of Figure 6.3 shows typical values of capacitances-to-ground of objects.

If the tank depicted in Figure 6.2 were isolated from the ground (i.e., $R = \infty$), the stored electric charge would continue to increase indefinitely. In this scenario, the ground potential $v(t)$ described by Equation (6.1) would rise linearly over time, as the exponential term would effectively become zero. This means that $v(t)$ would increase without bound as t increases. When the magnitude of $v(t)$ exceeds the dielectric strength of the air, a spark would occur, resulting in a discharge of the stored charge. This discharge momentarily neutralizes the accumulated charge, allowing the charging process to begin anew. This cycle

Figure 6.3: Capacitances-to-ground of objects.

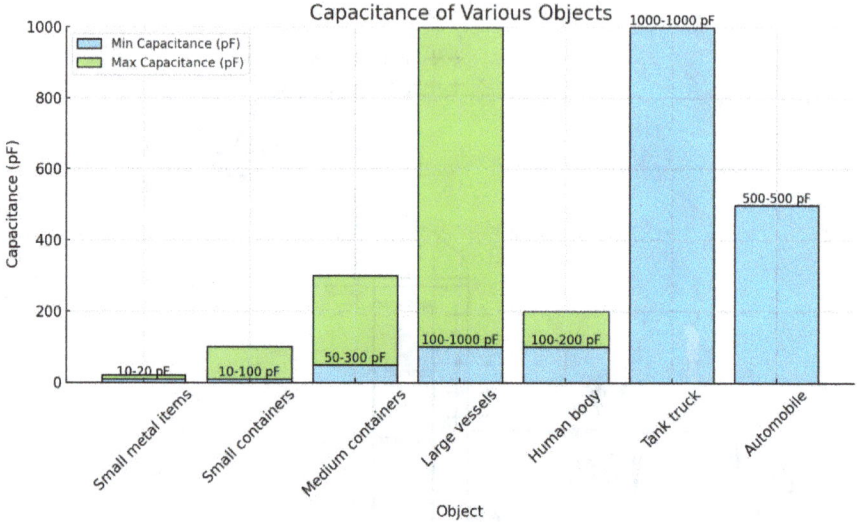

of charge accumulation and discharge would repeat continuously, governed by the charging current I and the dielectric properties of the surrounding medium.

Minimizing the relaxation time to permit a quicker dissipation of the static charge, reducing the risk of hazardous discharge events, such as sparks or explosions, is essential for ensuring safety. To achieve this, and effective practice is to enhance the electrical conductivity of liquids by incorporating a static dissipative additive. This way, the liquid can conduct charge more efficiently, facilitating faster dissipation and thus minimizing the relaxation time.

To prevent the build-up of static charge, it is also recommended to avoid using splash filling methods. This involves positioning the fill pipe above the liquid level in the tank, which causes the liquid to splash and create turbulence upon impact. Splash filling tends to generate static electricity due to the agitation and mixing of air and liquid. Instead, submerged filling methods should be employed. In submerged filling, the fill pipe extends close to the bottom of the tank, which reduces agitation and the formation of air bubbles, thereby minimizing static charge generation.

When dealing with large conductive tanks filled with low conductivity liquids, it is crucial to maintain the fluid flow velocity below 1 m s^{-1}. High flow velocities can induce turbulence, which accelerates the rate of charge separation and accumulation. To mitigate this risk, the pipe diameter to control the flow rate can be adjusted, ensuring that the flow remains laminar and minimizing static charge build-up.

6.4 Managing Stored Energy to Prevent Explosions in Hazardous Environments

The energy stored in a charged object is described by Equation (6.3):

$$W(t) = \frac{1}{2}Cv^2(t) = \frac{1}{2}CI^2R^2(1 - e^{-\frac{t}{RC}})^2. \tag{6.3}$$

If the charging process continues for a time of at least 3RC, the exponential term in Equation (6.3) can be disregarded, simplifying the expression for the stored energy. Under these conditions, the stored energy becomes similar to that of a capacitor charged to the voltage $V = IR$, where R is the resistance-to-ground of the object.

$$W \approx \frac{1}{2}C(IR)^2. \tag{6.4}$$

In environments where flammable gases, vapors, or dusts are mixed with air, possibly creating an ignitable mixture, the stored energy must be carefully managed to prevent explosions. If the stored energy of the object exceeds the minimum ignition energy W_{MIE} of the hazardous atmosphere, an explosion may occur. W_{MIE} is defined as the lowest amount of energy required to ignite the most easily ignitable mixture of a combustible substance (gas, vapor, or dust) with air. Therefore, to mitigate the risk of electrostatic discharge and associated hazards, it is crucial to calculate the maximum acceptable value R for the resistance-to-ground of the object. This ensures that the stored electrostatic energy does not exceed the minimum ignition energy of the surrounding atmosphere.

The following inequality can be used to determine the maximum permissible resistance R:

$$R < \frac{1}{I}\sqrt{\frac{2W_{MIE}}{C}}. \tag{6.5}$$

For most gases and vapors, the minimum ignition energy, typically ranges between 0.1 mJ and 0.3 mJ. In contrast, the minimum ignition energy for dusts can vary significantly, ranging from below 1 mJ to over 10 J.

To illustrate the potential risk, consider a large vessel situated in an explosive atmosphere with a concentration of methane. The W_{MIE} for methane is 0.21 mJ, serving as a critical threshold. If the stored electrostatic energy of the vessel, accumulated during filling operations with non-conductive liquids, exceeds this value, there is a significant risk of ignition. Using inequality (6.5), the maximum permissible resistance-to-ground R is calculated to be 9.2 MΩ. This indicates that even relatively high ground-resistance values are sufficient to prevent hazardous static charge accumulation and ensure that stored energy remains below the 0.21 mJ threshold. If the ground-resistance exceeds 9.2 MΩ, an effective grounding system must be implemented to mitigate the risk of electrostatic discharge and prevent ignition of the explosive methane-air mixture.

6.4.1 Simplified criterion to prevent charge accumulation based on breakdown potential

An incendive electrostatic discharge occurs when the electric field created by the stored charge of a conductive body exceeds the dielectric breakdown strength of the air, and the energy released during the discharge surpasses the minimum ignition energy of the surrounding flammable atmosphere. For typical industrial operations, the critical electrical potential threshold usually ranges from 300 V to 1 kV. Adopting a conservative approach and using 200 V as the critical potential, the maximum acceptable value for the resistance-to-ground R of a charged object for safe dissipation of static electricity can be calculated using Equation (6.6):

$$R = \frac{200 \text{ V}}{I} \tag{6.6}$$

where R is expressed in ohms and I, the charging current, in amperes, which can range from 10 pA to 100 μA. Consequently, the permissible values for the resistance-to-ground R can vary significantly, from 20 TΩ to 2 MΩ, respectively. For a maximum charging current of 100 μA, a resistance-to-ground not exceeding 2 MΩ would ensure safe dissipation of static electricity.

It is important to note that this calculation is a simplified approximation but provides a useful starting point for assessing and mitigating electrostatic hazards in industrial environments. Conservative guidelines in the literature suggest that a resistance-to-ground value of 1 MΩ or less generally ensures safe dissipation of static electricity in all situations. Therefore, conductive bodies with a resistance-to-ground exceeding 1 MΩ should be properly grounded to reduce the risks associated with static charge accumulation.

6.5 Grounding and Bonding

Grounding and bonding of conductive bodies are crucial strategies for mitigating static electricity hazards. Grounding prevents the accumulation of static charge and the dangerous release of stored electrostatic energy, while bonding maintains conductive parts at the same electrical potential to avoid potential differences that could lead to static discharge.

For example, Figure 6.4 illustrates the use of a grounding connection to allow any charge accumulated on the airplane while moving through the air to dissipate safely to the earth, ensuring secure and efficient refueling operations. The bonding connection prevents static discharge between the aircraft and the fuel tanker during refueling.

Figure 6.4: Grounding of an airplane and bonding to a fuel tanker.

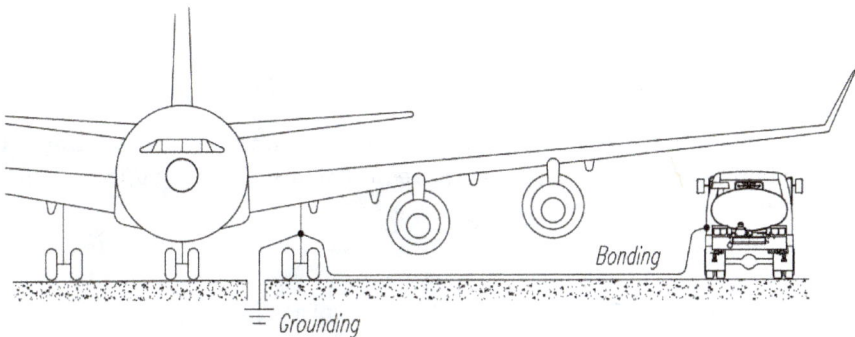

Another instance of bonding and grounding is shown in Figure 6.5, where a fuel tanker and an underground storage tank at a gas station are depicted.

Before unloading fuel, any charge present on the tank truck's body must be drained using an equipotential bonding connection to the tank or a ground-electrode with a resistance-to-ground of less than 1 MΩ. The pit where the connection is made is classified as Zone 1, indicating the presence of flammable substances mixed with air during normal operations and the potential for an explosive atmosphere. The connection to the tank could generate a spark, so the ground clamp must meet specific safety standards, such as an explosion-proof design (Ex d), suitability for use with flammable substances (IIA), and a surface temperature rating of 200 °C (T3). Additionally, the ground clamp should have

Figure 6.5: Grounding and bonding of tank truck.

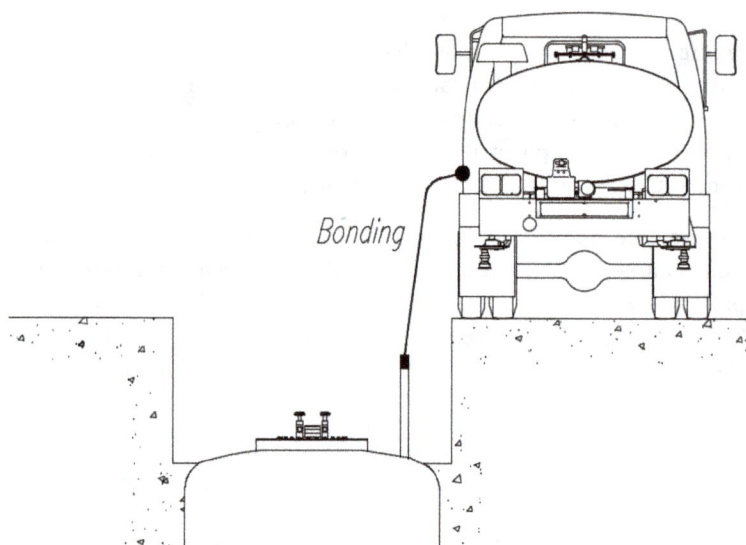

an insulated handle to prevent accidental discharge through the operator. If the clamping occurs outside the pit, the requirements for the clamp may be less stringent.

6.6 Static Charge Accumulation in the Human Body and Hazardous Discharge

The human body acts as an electrical conductor and can accumulate static charge when insulated from the ground. Various mechanisms contribute to static charge accumulation, such as walking across insulating flooring or removing and donning clothing. While discharges of this energy may sometimes go unnoticed, they can also cause sensations of pain or involuntary movements and falls. In environments where flammable gases, vapors, or combustible dusts are present, these discharges pose a significant hazard, particularly in locations with explosive atmospheres where the minimum ignition energy is less than 10 mJ.

For perspective, consider a person with a capacitance-to-ground of 150 pF walking across a carpet under dry conditions, in a room with a relative humidity ranging between 10% to 20%. In this scenario, the body's electrostatic voltage, due to triboelectric charging, can reach 35 kV (as indicated in NFPA 77[2]). Consequently, from Equation (6.4), the resulting electrostatic energy is approximately 91.8 mJ, which is sufficient to ignite many explosive atmospheres.

To mitigate the risks associated with static charge accumulation and discharge in the human body, several safety measures can be implemented. Installing conductive or static-dissipative flooring materials can help prevent the build-up of static electricity by facilitating charges to flow to the ground. Wearing conductive or static-dissipative footwear ensures that any charge accumulated on the body is safely dissipated to the ground through the floor. Maintaining a higher humidity level in the environment can reduce static charge build-up, as moisture in the air increases the conductivity of surfaces and materials. Using anti-static clothing made of materials that do not easily accumulate static charge can reduce the risk of static discharge. These materials are often used in environments where the handling of sensitive electronic components or flammable materials is common. In certain industrial or laboratory settings, grounding wrist straps can be used to continuously discharge any accumulated static charge to the ground, ensuring that workers remain at a safe potential.

6.6.1 Static Dissipative Floor and Footwear

The optimal resistance-to-ground value for static dissipative flooring systems is identified in technical standards to be less than 100 MΩ. A floor with such resistance-to-ground allows for the dissipation of static charges, while still being high enough to limit electrical shock hazards. As previously discussed, floors are considered insulating if their resistance-to-ground is at least 50 kΩ for systems with voltages up to 500 V and 100 kΩ for locations operating at voltages exceeding 500 V. As a reference, clean uncoated concrete floors typically have a resistance-to-ground ranging between 1 MΩ and 100 MΩ. However, accumulation of debris or other high-resistivity materials can compromise the floor conductivity, and therefore prevent the charge dissipation. Regular inspections and resistance measurements should be regularly conducted to ensure the floor's resistance remains within the desired range. Any necessary cleaning or maintenance should be promptly addressed to maintain the effectiveness of the static dissipative properties.

In addition to flooring, static dissipative footwear plays a crucial role in preventing the accumulation of static charge on the human body. These shoes are designed to provide a conductive path from the body to the floor, ensuring that

---

[2]NFPA 77 "Recommended Practice on Static Electricity."

any static charge accumulated while walking is safely dissipated. Dissipative footwear features resistance levels that can still provide adequate resistance to offer protection against electric shock.

Various factors can cause an unwanted increase in footwear resistance, making regular testing and monitoring essential. These factors include the accumulation of debris or non-conductive deposits on the soles, the use of orthopedic insoles, and a decrease in the contact area with the floor. Generally, socks do not affect the properties of dissipative or conductive footwear.

Combining static dissipative flooring and footwear creates a comprehensive system that effectively minimizes the risk of static discharge, particularly in environments where the presence of flammable gases, vapors, or combustible dusts poses significant hazards.

6.7 Summary

This chapter delves into the fundamental aspects of static electricity, including its generation, accumulation, and the potential hazards it poses in various environments. Static charge is generated primarily through the separation of positive and negative charges between dissimilar materials, a process known as triboelectric charging. This charge imbalance results from the transfer of electrons when materials come into contact and then separate. Key factors influencing static charge production include material properties, surface area, and environmental conditions.

Materials exhibit varying tendencies to gain or lose electrons, which is defined by their position in the triboelectric series. Surface area and the nature of the surface also affect the efficiency of charge transfer. Environmental factors like humidity and temperature further impact static charge dynamics, with higher humidity generally reducing charge accumulation.

Static charge can accumulate on conductive, insulating, and high-resistivity objects, leading to potential differences that can reach up to 100 kV. To mitigate the risks associated with high static charge, grounding is essential. This practice prevents charge accumulation and reduces the risk of hazardous discharges such as sparks, which could ignite flammable substances.

The chapter also explores the classification of materials based on their electrical resistivity, as per IEC standards. Materials are categorized as conductive, dissipative, or insulating, with each type having distinct characteristics that affect static charge dissipation. For instance, insulating materials retain electric charges due to high resistivity, while conductive and dissipative materials facilitate charge dissipation.

In the context of liquids, static charge can accumulate due to friction during flow, especially in low-conductivity liquids. Grounding containers and employing proper filling techniques can mitigate these risks. Additionally, the chapter discusses the impact of relaxation time on static charge dissipation, emphasizing the importance of enhancing electrical conductivity to reduce potential hazards.

For environments with flammable substances, managing stored electrostatic energy is crucial. The chapter provides formulas to calculate the maximum resistance-to-ground to ensure stored energy does not exceed the minimum ignition energy of the surrounding atmosphere. It also discusses simplified criteria based on breakdown potential to prevent incendiary electrostatic discharges.

Grounding and bonding are highlighted as effective strategies for static electricity control, with examples from industrial operations such as fuel refueling. Proper grounding and bonding prevent the accumulation and discharge of static energy, ensuring safe operations.

Finally, the chapter addresses the risks associated with static charge accumulation in the human body. It outlines preventive measures, including the use of static dissipative flooring and footwear, maintaining appropriate humidity levels, and employing anti-static clothing. These measures are essential for reducing static discharge risks, particularly in environments where explosive atmospheres are present. Regular monitoring and maintenance of static dissipative systems are recommended to ensure ongoing effectiveness.

Overall, this chapter provides a comprehensive overview of static electricity management, emphasizing the importance of understanding static charge dynamics and implementing effective safety measures to mitigate associated hazards.

Key Definitions

Antistatic materials: Substances that are designed to reduce or prevent the build-up of static electricity by either conducting or dissipating charge.

Capacitance-to-ground: The measure of a conductive object's ability to store charge relative to the ground. It represents how easily the object can accumulate static charge when isolated from the ground.

Charging current: The rate at which electric charge accumulates on an object. This current influences the buildup of static charge over time.

Conductive materials: Materials with low electrical resistance that allow the easy flow of electric charge. Examples include metals like copper and aluminum.

Dissipative materials: Materials that have intermediate electrical resistance and can conduct static electricity, thereby helping to dissipate charges.

Electrostatic discharge: The release of static electricity when two objects at different potentials come into contact or proximity, potentially igniting flammable materials.

Grounding: The process of connecting a conductive object to ground to allow static charge to safely dissipate, preventing the build-up of hazardous levels of static electricity.

Insulating materials: Materials with high electrical resistance that impede the flow of electric charge, often used to prevent the transfer of static charge. Examples include rubber and glass.

Relaxation time (τ): The time required for the static charge on a surface or within a bulk material to decrease to 37% of its initial value after charge generation has ceased. It depends on the material's conductivity.

Static charge: An imbalance of electric charge on the surface of an object, resulting from the transfer of electrons between materials during contact and separation.

Static dissipative footwear: Footwear designed to safely conduct static electricity from the human body to the ground, reducing the risk of static charge accumulation.

Static dissipative flooring: Flooring designed to control static charge by providing a conductive path to ground, thereby reducing the risk of static electricity buildup.

Triboelectric charging: The generation of static electricity through the contact and separation of two different materials, leading to a transfer of electrons and resulting charge imbalance.

Voltage-to-ground: The electric potential difference between a charged object and the ground, which influences the risk of electrostatic discharge.

Wearable anti-static devices: Personal protective equipment, such as wrist straps or grounding bracelets, designed to prevent the accumulation of static charge on individuals working in sensitive environments.

Bibliography

1. M. Mitolo: "Electrical Safety of Low-Voltage Systems," McGraw-Hill, 2009. ISBN: 978-0071508186.
2. IEC 60079-0:2017 "Explosive atmospheres - Part 0: Equipment - General requirements."
3. ISO 80079-37:2016 "Explosive atmospheres Part 37: Non-electrical equipment for explosive atmospheres."
4. ISO 12100:2010 "Safety of machinery General principles for design. Risk assessment and risk reduction."
5. H. S. Silva, D. Fernandes Leite Pereira, E. V. Falcão, H. E. Querino de Carvalho and V. V. Freire, "Electrostatic energy: A possible source of interference," 2015 IEEE International Conference on Microwaves, Communications, Antennas and Electronic Systems (COMCAS), Tel Aviv, Israel, 2015.
6. IEC TS 60079-32-1:2017 "Explosive atmospheres – Part 32-1: Electrostatic hazards, guidance."
7. NFPA 77:2024 "Recommended Practice on Static Electricity."
8. M. Mitolo, F. Freschi, R. Tommasini, "Analysis of causation of a flour dust explosion in an industrial plant," in IEEE Transactions on Industry Applications Vol. 53, No. 6; November/December 2017.
9. M. Mitolo, "Protective Bonding Conductors: An IEC Point of View," in IEEE Transactions on Industry Applications, vol. 44, no. 5, Sept.-Oct. 2008.
10. IEC 60079-10-1:2020, "Explosive atmospheres - Part 10-1: Classification of areas - Explosive gas atmospheres."
11. IEC 60079-0:2017, "Explosive atmospheres - Part 0: Equipment - General requirements."
12. T. Viheriäkoski, M. Kokkonen, P. Tamminen, E. Kärjä, J. Hillberg and J. Smallwood, "Electrostatic threats in hospital environment," Electrical Overstress/Electrostatic Discharge Symposium Proceedings 2014, Tucson, AZ, USA, 2014.
13. Italian standard CEI 64-8/2, "Low-voltage electrical installations."
14. IEC 60364-6:2016, "Low voltage electrical installations - Part 6: Verification."
15. IEC 61340-4-1:2003+AMD1:2015, "Electrostatics - Part 4-1: Standard test methods for specific applications - Electrical resistance of floor coverings and installed floors."
16. IEC 61340-4-5 Ed. 1.0 b:2004, "Electrostatics - Part 4-5: Standard Test Methods for Specific Applications - Methods for Characterizing the Electrostatic Protection of Footwear and Flooring In Combination With A Person."
17. M. Mitolo, et al. "Electrostatic Hazards in Power Systems," I&CPS Europe 2024.

Index

For Product Safety Concerns and Information please contact our EU
representative GPSR@taylorandfrancis.com
Taylor & Francis Verlag GmbH, Kaufingerstraße 24, 80331 München, Germany

www.ingramcontent.com/pod-product-compliance
Lightning Source LLC
Chambersburg PA
CBHW061611220326
41598CB00024BC/3536